ALSO BY EMANUEL DERMAN

*My Life as a Quant: Reflections on Physics and Finance*

# MODELS.BEHAVING.BADLY.

Why Confusing Illusion with Reality
Can Lead to Disaster, on Wall Street and in Life

## EMANUEL DERMAN

FREE PRESS

New York   London   Toronto   Sydney   New Delhi

FREE PRESS
A Division of Simon & Schuster, Inc.
1230 Avenue of the Americas
New York, NY 10020

First Free Press hardcover edition October 2011

FREE PRESS and colophon are trademarks of Simon & Schuster, Inc.

For information about special discounts for bulk purchases, please contact
Simon & Schuster Special Sales at 1-866-506-1949 or business@simonandschuster.com.

The Simon & Schuster Speakers Bureau can bring authors to your live event.
For more information or to book an event contact the Simon & Schuster Speakers Bureau
at 1-866-248-3049 or visit our website at www.simonspeakers.com.

*Grateful acknowledgment is made to the following for permission to reprint
previously published material:*

"The Precision of Pain and the Blurriness of Joy: The Touch of Longing Is Everywhere"
from *Open Closed Open* copyright © 2000 by Yehuda Amichai, English translation
copyright © 2000 by Chana Bloch and Chana Kronfeld, reprinted by permission of
Houghton Mifflin Harcourt Publishing Company.

John Maynard Keynes, *Essays in Biography*, published 2010,
reproduced with permission of Palgrave Macmillan.

"This Be the Verse" from *Collected Poems* by Philip Larkin. Copyright © 1988, 2003 by the
Estate of Philip Larkin. Reprinted by permission of Farrar, Straus and Giroux, LLC.

DESIGNED BY ERICH HOBBING

Manufactured in the United States of America

5   7   9   10   8   6   4

Library of Congress Cataloging-in-Publication Data

Derman, Emanuel.
Models.behaving.badly. : why confusing illusion with reality can lead to disaster,
on Wall Street and in life / Emanuel Derman.
p. cm.
Includes bibliographical references.
1. Mathematical models—Social aspects. 2. Theory (Philosophy)—Social aspects.
3. Metaphor—Social aspects. 4. Derman, Emanuel. I. Title.
QA401.D37 2011
003—dc22    2011015006

ISBN 978-1-4391-6498-3
ISBN 978-1-4391-6501-0 (ebook)

# CONTENTS

## I. MODELS

## II. MODELS BEHAVING

## III. MODELS BEHAVING BADLY

# MODELS.BEHAVING.BADLY.

# I. MODELS

# CHAPTER 1

# A FOOLISH CONSISTENCY

*Models that failed • Capitalism and the great financial crisis • Divining the future via models, theories, and intuition • Time causes desire • Disappointment is inevitable • To be disappointed requires time, desire, and a model • Living under apartheid • Growing up in "the movement" • Tat tvam asi*

> Pragmatism always beats principles. . . . Comedy is what you get when principles bump into reality.
>
> —J. M. Coetzee, *Summertime*

## MODELS THAT FAILED I: ECONOMICS

"All that is solid melts into air, all that is holy is profaned, and man is at last compelled to face, with sober senses, his real conditions of life, and his relations with his kind," wrote Marx and Engels in *The Communist Manifesto* in 1848. They were referring to modern capitalism, a way of life in which all the standards of the past are supposedly subservient to the goal of efficient, timely production.

With the phrase "melts into air" Marx and Engels were evoking sublimation, the chemists' name for the process by which a solid transmutes directly into a gas without passing through an intermediate liquid phase. They used sublimation as a metaphor to describe the way capitalism's endless urge for new sources of profits results in the destruction of traditional values. Solid-to-vapor is an apt summary of the evanescence of value, financial and ethical, that has taken place throughout the great and ongoing financial crisis that commenced in 2007.

The United States, the global evangelist for the benefits of creative destruction, has favored its own church. When governments of emerging markets complained that foreign investors were fearfully yanking capital from their markets during the Asian financial crisis of 1997, liberal democrats in the West told them that this was the way free markets worked. Now we prop up our own markets because it suits us to do so.

The great financial crisis has been marked by the failure of models both qualitative and quantitative. During the past two decades the United States has suffered the decline of manufacturing; the ballooning of the financial sector; that sector's capture of the regulatory system; ceaseless stimulus whenever the economy has wavered; taxpayer-funded bailouts of large capitalist corporations; crony capitalism; private profits and public losses; the redemption of the rich and powerful by the poor and weak; companies that shorted stock for a living being legally protected from the shorting of their own stock; compromised yet unpunished ratings agencies; government policies that tried to cure insolvency by branding it as illiquidity; and, on the quantitative side, the widespread use of obviously poor quantitative security valuation models for the purpose of marketing.

People and models and theories have been behaving badly, and there has been a frantic attempt to prevent loss, to restore the status quo ante at all costs.

## THEORIES, MODELS, AND INTUITION

For better or worse, humans worry about what's ahead. Deep inside, everyone recognizes that the purpose of building models and creating theories is divination: foretelling the future, and controlling it.

When I began to study physics at university and first experienced the joy and power of using my mind to understand matter, I was fatally attracted. I spent the first part of my professional life doing research in elementary particle physics, a field whose theories are capable of making predictions so accurate as to defy belief. I spent the second part as a professional analyst and participant in financial markets, a field in which sophisticated but often ill-founded models abound. And all the while I observed myself and the people around me and the assumptions we made in dealing with our lives.

What makes a model or theory good or bad? In physics it's fairly easy to tell the crackpots from the experts by the content of their writings, without having to know their academic pedigrees. In finance it's not easy at all. Sometimes it looks as though anything goes. Anyone who intends to rely on theories or models must first understand how they work and what their limits are. Yet few people have the practical experience to understand those limits or whence they originate. In the wake of the financial crisis naïve extremists want to do away with financial models completely, imagining that humans can proceed on purely empirical grounds. Conversely, naïve idealists pin their faith on the belief that somewhere just offstage there is a model that will capture the nuances of markets, a model that will do away with the need for common sense. The truth is somewhere in between.

In this book I will argue that there are three distinct ways of understanding the world: theories, models, and intuition. This book is about these modes and the distinctions and overlaps between them. Widespread shock at the failure of quantitative models in the mortgage crisis of 2007 results from a misunderstanding of the difference between models and theories. Though their syntax is often similar, their semantics is very different.

*Theories* are attempts to discover the principles that drive the world; they need confirmation, but no justification for their existence. Theories describe and deal with the world on its own terms and must stand on their own two feet. *Models* stand on someone else's feet. They are metaphors that compare the object of their attention to something else that it resembles. Resemblance is always partial, and so models necessarily simplify things and reduce the dimensions of the world. Models try to squeeze the blooming, buzzing confusion into a miniature Joseph Cornell box, and then, if it more or less fits, assume that the box is the world itself. In a nutshell, theories tell you what something is; models tell you merely what something is like.

*Intuition* is more comprehensive. It unifies the subject with the object, the understander with the understood, the archer with the bow. Intuition isn't easy to come by, but is the result of arduous struggle.

What can we reasonably expect from theories and models, and why? This book explains why some theories behave astonishingly well, while some models behave very badly, and it suggests methods for coping with this bad behavior.

## OF TIME AND DESIRE

In "Ducks' Ditty," the little song composed by Rat in Kenneth Grahame's *The Wind in the Willows*, Rat sings of the ducks' carefree pond life:

> *Everyone for what he likes!*
> *We like to be*

# A FOOLISH CONSISTENCY

*Heads down, tails up,*
*Dabbling free!*

Doubtless the best way to live is in the present, head down and tail up, looking at what's right in front of you. Yet our nature is to desire, and then to plan to fulfill those desires. As long as we give in to the planning, we try to understand the world and its evolution by theories and models. If the world were stationary, if time didn't pass and nothing changed, there would be no desire and no need to plan. Theories and models are attempts to eliminate time and its consequences, to make the world invariant, so that present and future become one. We need models and theories because of time.

Like most people, when I was young I couldn't imagine that life wouldn't live up to my desires. Once, watching a TV dramatization of Chekhov's "Lady with a Lapdog," I was irritated at the obtuse ending. Why, if Dmitri Gurov and Anna Sergeyevna were so in love, didn't they simply divorce their spouses and go off with each other?

Years later I bought a copy of Schopenhauer's *Essays and Aphorisms*. There I read an eloquent description of time's weary way of dealing with human aspirations. In his 1850 essay "On the Suffering of the World" Schopenhauer wrote:

> If two men who were friends in their youth meet again when they are old, after being separated for a life-time, the chief feeling they will have at the sight of each other will be one of complete disappointment at life as a whole, because their thoughts will be carried back to that earlier time when life seemed so fair as it lay spread out before them in the rosy light of dawn, promised so much—and then performed so little. This feeling will so completely predominate over every other that they will not even consider it necessary to give it words, but on either side it will be silently assumed, and form the ground-work of all they have to talk about.

Schopenhauer believed that both mind and matter are manifestations of the Will, his name for the substance of which all things are made, that thing-in-itself whose blind and only desire is to endure. Both the world outside us and we ourselves are made of it. But though we experience other objects from the outside as mere matter, we experience ourselves from both outside and inside, as flesh *and* soul. In matter external to us, the Will manifests itself in resilience. In our own flesh, the Will subjects us to endless and unquenchable desires that, fulfilled or unfulfilled, inevitably lead to disappointments over time.

You can be disappointed only if you had hoped and desired. To have hoped means to have had preconceptions—models, in short— for how the world should evolve. To have had preconceptions means to have expected a particular future. To be disappointed therefore requires time, desire, and a model.

I want to begin by recounting my earliest experiences with models that disappoint.


## MODELS THAT FAILED II: POLITICS

I grew up in Cape Town, South Africa, in a society where most white people had Coloured servants, sometimes even several of them. Their maids or "boys" lived in miserably small rooms attached to the outside of the "master's" house. Early in my childhood the Afrikaner Nationalist Party government that had just come to power passed the Prohibition of Mixed Marriages Act of 1949. The name speaks for itself. Next came the Immorality Act of 1950, which prohibited not just marriage but also adultery, attempted adultery, and other "immoral" acts between whites and blacks, thereby trying to deny, annul, or undo 300 years of the miscegenation that was flagrantly visible. In South Africa there were millions of "Cape Coloureds," people of mixed European and African ancestry, who lived in the southern part of the country, their skin tone ranging from indistinguishable-from-white to indistinguishable-from-black and including everything in between.

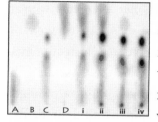

In South Africa we all became expert at a social version of chromatography, a technique chemists use to separate the colors within a mixture. I learned how to do it in my freshman chemistry course at the University of Cape Town. You place a drop of black ink on a strip of blotting paper and then dip the end of the strip into water. As the water seeps through the paper, it transports each of the different dyes that compose black through a different distance, and, as if by magic, you can see the colors separate. How convenient it would have been for the government to put each person into a device that could have reported his or her racial composition scientifically. But the authorities came as close to that as they could: the Population Registration Act of 1950 created a catalogue in which every individual's race was recorded. South Africa didn't just categorize people into simple black and white; there were whites, natives (blacks), Coloureds, and Indians. Racial classification was a tortuous attempt to impose a flawed model on unruly reality:

A white person is one who in appearance is, or who is generally accepted as, a white person, but does not include a person who, although in appearance obviously a white person, is generally accepted as a Coloured person.

A native is a person who is in fact or is generally accepted as a member of any aboriginal race or tribe of Africa.

A Coloured person is a person who is not a white person nor a native.

Note the pragmatic combination of objectivity and subjectivity: if you are objectively white but accepted as Coloured, then you're not white.

In disputed cases a board made decisions that determined not only who you could sleep with but which beaches you could swim at, where you could work and live, which buses you could take, and which cinemas you could attend. Given South Africa's history of miscegenation, it was not uncommon for members of the same family to end up with different chromatography profiles. Some Coloureds attempted to be reclassified as white, and some blacks applied to be reclassified as Coloured. Evidence involved keen discussions of texture of bodily hair, nose shape, diet, and ways of earning a living, the latter two being taken as racial characteristics rather than matters of socialization or opportunity. Most Chinese, who were difficult for officials to define or even to distinguish from other Asians, were classified as nonwhite, but Chinese from Taiwan and all Japanese, for trade and economic reasons, were declared honorary whites.

The Group Areas Act of 1950 institutionalized apartheid by specifying the regions in which each race could live and do business. Nonwhites were forcibly removed from living in the "wrong" areas, thereby superimposing a legal separation over the less formal physical separation of the races that had already existed.[1] Those domestics who didn't "live in" had to commute long distances to work. In Cape Town the government razed District 6, its Coloured Harlem, and moved the entire community of inhabitants to the Cape Flats, a desolate sandy region outside the city, well described by its name. When I was at university I trekked out there several times as a volunteer on behalf of the Cape Flats Development Association to help persuade poor Coloured families to feed their children milk rather than the cheaper mashed-up squash that, though stomach-filling, had virtually no nutritional value. It was a bleak area with sparse vegetation and no running water, a gulag whose inhabitants lived in makeshift shanties constructed of corrugated iron, plywood, and cardboard. Barefoot children were everywhere. Many parts of South Africa are still like that, despite the end of apartheid.

By 1951 nonwhites were being stripped of whatever voting rights they had possessed. Though I knew all this was wrong, I grew up with it as normality. The air you breathe, once you grow accustomed to it, has no smell at all.[2]

When I was ten years old our neighbor down the block, a Jewish businessman in his forties with two sons a little older than I, was found on the floor of his downtown office in flagrante delicto with a young black girl. His doctor testified that he had prescribed pills for our neighbor's heart condition that might have had aphrodisiac side effects. The black girl apparently didn't need pills to provoke her desire, and I don't recall what sentence, if any, either of them received.

Several years later an acquaintance of my sister's was arrested. The police had seen him driving in his car at night with a Coloured woman seated beside him. They trailed him to his house, watched through the window, and later testified to observing the sexual act. His stained underwear was presented in court as evidence. The initial giveaway was the fact that the woman sat in the front seat, beside him. White men who gave their maids a ride somewhere commonly made them sit in the backseat to avoid suspicion.

But even white women (the "madams") often made their maids sit in the backseat. The unarticulated aim was the avoidance of even innocuous physical intimacy. (Of course, if it had to be avoided, it wasn't innocuous.) A native's lack of whiteness made him or her untouchable. To avoid contamination, white families often had two sets of knives, forks, and plates: one for the family to use and one for their maids and "boys." When I read *Portnoy's Complaint* in 1969, a few years after I arrived in New York, the following passage reminded me of the visceral sense of defilement that many South African whites had been taught to feel:

11

Once Dorothy chanced to come back into the kitchen while my mother was still standing over the faucet marked H, sending torrents down upon the knife and fork that had passed between the schvartze's thick pink lips. "Oh, you know how hard it is to get mayonnaise off silverware these days, Dorothy," says my nimble-minded mother— and thus, she tells me later, by her quick thinking has managed to spare the colored woman's feelings.

———

The Nationalist Party government that came to power in 1948 hated and feared Communism, not because the Nationalists were lovers of the individual freedom threatened by totalitarianism, but because they were totalitarian themselves. They denounced "radicals," but as a student leader at a University of Cape Town rally once pointed out to great applause, it was the Nationalists who were the true radicals, intent on wiping out age-old conservative democratic principles. Their government periodically declared a state of emergency, which allowed for arbitrary detention. They put opponents and suspects in jail without trial for 180 days, renewable. Eventually they banned the Communist Party. Then they proceeded to ban the more gentlemanly Liberal Party, whose slogan was "One man, one vote." Fearful people made an effort to say they were "liberal with a small l."

When I was seventeen and spending the summer working and touring in Israel, I bought a copy of *Atlas Shrugged* and hid it in my luggage on my return, successfully slipping it through Customs like a copy of *Playboy* or *Tropic of Cancer*. The South African prism had shifted the political spectrum so far to the dictatorial right that Ayn Rand's defense of the individual and of libertarian capitalism seemed to me and my friends to be subversive. At the extremes, left could not be distinguished from right. I thought of this later, when I first learned the theory of complex numbers: in the complex plane, the points at plus and minus infinity coincide, and again far left and far right become indistinguishable.

South Africa's models were rife with internal contradictions. The most severe was the government's policy of race separation that pretended to grant blacks independence in their supposed homelands while still keeping them available to provide the labor that kept the country running. There were smaller hypocrisies too. As young white teenagers in the 1950s, we spent the entire summer in the sun on Fourth Beach at Clifton or in the crowded Snake Pit at Muizenberg, applying fish oil or Skol so as to get as dark as possible.[3] A girl I knew who devoted her time to acquiring a magnificent tan grew indignant when the train conductor mistook her for a Coloured and instructed her to go to the train carriage reserved for that race.

Coloureds were treated better than natives but much worse than whites. Their facilities weren't separate but equal; they were vastly inferior or nonexistent. In downtown Cape Town, where I worked in a department store one summer in the early 1960s, I don't think there was a single restaurant a black person could enter to sit down and eat. All the salesladies behind the counter, even in down-market OK Bazaars, were white.

From birth I knew no other society, and though I knew apartheid was wrong, individual blacks were pretty much invisible to me. Once, soon after I learned to drive, I took my parents' car to the garage to get petrol. In those distant days of luxury all garages were full service, and the "boys" bustled around your car when you drove up. They pumped petrol; checked the oil, water, battery, and brake and clutch fluids; cleaned the windows; and measured the tires' pressure and put in air if necessary. When you left, you tipped the attendant who had served you. That day, my nervous first time dealing with a garage on my own, there were three or four attendants hovering around the several cars at the petrol pumps, and as I drove away I realized with minor horror that I had mistakenly tipped the wrong man. When you weren't used to seeing blacks as individuals, they truly did all look the same.

Enforced racial separation hadn't always been the norm. I spent my first seven years in Salt River, a poor mixed-race suburb that was home to many immigrant Jews who hadn't yet made it. (I remember fondly Mr. Jenkins, our Coloured plumber, who lived in the neighborhood. He spoke Yiddish, and once, when he arrived at our front door while I was in bed with a bad cold, I fearfully mistook his voice and intonation for that of our doctor, who also made home visits.) Apartheid as a legal policy reached peak efficiency only in the late 1950s and 1960s, my formative years, when I became accustomed to racism. My sisters, 9 and 12 years older than I, grew up in a less formally prejudicial world and were less racist than I was. My nephews and nieces, 16 or more years younger, grew up as the apartheid regime was collapsing, and it left a milder indentation on them.

It was only when I left to study in New York in the late 1960s that I had the chance to socialize informally with people that South Africa classified as nonwhites. One day, kidding around physically with some Indian friends in the common room of the graduate student dormitory where we all lived, I suddenly realized that I was doing what I'd never done before, and was grateful for it.

---

When I was ten I spent the winter vacation with my parents about 100 miles northeast of Cape Town, in Montagu, a small town reached by steep switchbacks that crossed a deep ravine called DuToit's Kloof. Founded by British settlers in the mid-1800s, Montagu was a faded winter retreat, a Jewish immigrant's colonial-style Bath or Évian, but with a local population of Coloureds and Afrikaners. The town's main attraction was a nearby thermal spring that was reputedly good for arthritis. The refined hotel on the main street was called The Avalon. We stayed in The Baths, set in the countryside a few miles out of town. The Baths was fun but run-down. There was one toilet and bathroom at the end of each wing, and because it was a long, cold walk down the outdoor passage that connected the rooms, there was a heavy white enamel chamber pot beneath your bed in case you needed to urinate during the night. The Coloured maids emptied it in the morning, when they made up the room.

Baboons roamed the small kloof that separated The Baths from the business center of tiny Montagu. Sometimes they came onto the hotel grounds, emptying trash cans and even entering rooms. An older boy I knew climbed the hills above the hotel to shoot the baboons with an air gun, which I coveted.

The adults used to take a constitutional every morning, hiking into town through the kloof to The Avalon, to take tea and Scottish scones with local strawberry jam, butter, and thick whipped cream, but we children stuck to the grounds of The Baths, furiously socializing. My father babied me whenever I allowed him to and embarrassed me by forcing apples on me while I was with my friends. I fell in love with a twelve-year-old girl who scorned me, thanks to my father's constant attention. It was in Montagu that someone, I don't recall who, explained to me where babies come from. And it was in Montagu a few years later that I briefly met Adrian Leftwich.

———

Each year seasonal crazes swept through our school. One month it was silkworms that we bought and collected, keeping them in shoeboxes with airholes and feeding them mulberry or cabbage leaves until they grew into fabric-wrapped armatures. A season later came marbles. And then, outdoing all previous crazes, came hypnosis.

The sovereign of hypnosis in Cape Town was Max Collie, a professional entertainment hypnotist who had emigrated to South Africa from Scotland. His son and I went to the same school. Every year or so Mr. Collie did a couple of shows in Cape Town, some of them on our school's premises. He began by testing the audience for suggestibility, attempting to talk their outstretched right arms into floating up into the air while their eyes were closed.[4] "Your arm wants to rise up into the air. It feels light, like a balloon, so light it wants to float up towards the ceiling. Don't resist, let it go, let it go." Occasionally some hypersuggestible soul whose arm had spontaneously risen up would already be in a trance as a result of the test, and would fail to open his eyes at its conclusion, even before he had been officially hypnotized. Those suggestibles who were uninhibited enough to agree to participate in the show then went onstage to be hypnotized in front of the entire audience, including their own children. Soon adult men and women were under Mr. Collie's command, shyly attending their first day at school, asking the teacher for permission to go to the washroom, scratching as though there were itching powder in their clothes, lying rigidly across two separated chairs. Finally, there was the post-hypnotic suggestion: "When you wake up and are back in the audience, whenever you hear me say 'It is very warm in here tonight,' you will feel as though you are sitting on a hot electric plate and jump up screaming." Then he woke them: "As I count backwards from ten to one, you will slowly start to feel wider and wider awake. Ten, nine, eight . . . you feel light and cheerful, your eyes are beginning to open . . . seven, six, five, four . . . you are almost ready to wake up, you feel very good and full of energy . . . three, two, one, wake up! Thank you very much, ladies and gentlemen."

It was awe-inspiring to see people under Max Collie's power, and soon we were all trying to hypnotize each other. I bought books on hypnosis and self-hypnosis written by the aptly named Melvin Powers. The covers had mesmerizing diagrams of vertigo-inducing centripetal spirals, and some of the books included "the amazing hypnodisk," which you could use to hypnotize yourself and your friends. My cousin and I spent hours trying to put each other under.[5]

In Montagu that winter of the hypnosis craze I first met the equally aptly named Adrian Leftwich, several years older than the rest of us and not really a part of our more childish circle. I didn't see him again until a few years later, in the early 1960s, when I went to the University of Cape Town. By then Leftwich was the charismatic head of the National Union of South African Students, or Nusas, a principled anti-apartheid group. He was one in a series of Nusas student leaders who were in outspoken opposition to the government, and I admired his leadership and courage. And it truly did take courage: many student leaders of Nusas, like other foes of apartheid whom the government despised and even feared, were frequently arrested and eventually "banned," legally forbidden to attend any public meetings or even go to the cinema or theater. A more extreme punishment was house arrest. Most of the banned had had their passports revoked, so if they chose to leave the country they had to do so on a one-time permit into permanent exile. Anti-apartheid rallies were monitored by policemen and plainclothes agents of the Special Branch, who took photographs, and even those who merely signed anti-apartheid petitions worried about getting their names on a blacklist.

As the government clamped down on all forms of legal protest, violent opposition emerged. In 1963 there were sabotage attacks on power pylons and FM transmitters in the vicinity of Cape Town. In 1964 the security police carried out nighttime searches of the houses of known anti-apartheid activists, Leftwich among them. They found him in bed with his girlfriend, his flat carelessly filled with detailed plans that incriminated him as the hitherto anonymous leader of the African Resistance Movement, which had taken responsibility for the sabotage. The police arrested Leftwich and kept him in solitary confinement. Perhaps fearful of being sentenced to death, he quickly turned state's evidence and, in his own words in a later written reminiscence, "named the names" of his collaborators and recruits and gave testimony for the prosecution at their trial. I attended court on the day of the sentencing, where the presiding judge said that to call Leftwich a rat would be an insult to the genus *Rattus*.

I never had much political courage and had admired Leftwich for his bravery as head of Nusas. I don't judge him now. Like most of us, he wasn't what he thought he was. But thankfully, for most of us, comprehension of the disparity between who we think we are and who we truly are comes gradually and with age. We are lucky to avoid a sudden tear in our self-image and suffer more easily its slow degradation. For Leftwich the apparent union between personality and character ruptured like the fuselage of the early De Havilland Comet, in an instant, in midair, unable to withstand the mismatch between external and internal pressure. How do you ever forgive yourself for a betrayal like that?

But we have all committed acts that surprise us and are hard to forgive. You can count yourself lucky if your model of yourself survives its collision with time.

## MODELS THAT FAILED III: THE MOVEMENT

I was the accidentally conceived last child of Jewish parents who emigrated from Poland (now Belarus) to Cape Town in the mid-1930s to get away from what they saw as the anti-Semitic Poles. My parents' departure from Poland turned out to be a fortuitous escape from the concentration camps, but my maternal grandparents and many of the uncles and aunts I never knew stayed behind and weren't as fortunate. Had my mother been certain her father was dead by 1945, I would have been named Nahum Zvi. Sixteen years later, in Jewish tradition, my nephew was given his name.

When I was four years old, in late 1949, our family took a six-week trip to Israel. My mother hadn't seen her only two surviving sisters and one brother since 1935, when she had embarked for South Africa and they had emigrated to Palestine. We took a propeller-driven DC Skymaster from Cape Town to Lydda Airport in Israel, stopping in

Johannesburg, Nairobi, Entebbe, Juba, Khartoum, Wadi Halfa, Cyprus, and several other places I don't now recall. An enormously fat man on our plane had a heart attack after eating some pickled meat somewhere over the Sudan. Officials met us on the tarmac when we next touched down, escorted us into the shade of a shack, and took him away. We had left summer behind in Cape Town; in Israel it was the now famously cold winter of 1949–1950. It snowed in Tel Aviv that year—it hasn't happened since—and unprepared for the severity of the cold, we wore pajamas underneath our clothes all day long. It was the aftermath of the Israeli War of Independence, and food was being rationed. I recall going with my aunt to the coupon bureau, where she pleaded for an extra banana for me. I remember everything quite vividly, the rooster-shaped red lollipops they sold in the stores, the corn on the cob scooped out of steaming pots by street vendors, the grapefruit my sister and cousin and I stole off the trees of an orchard. I remember too the blood-red eyeballs of my little Israeli cousin, two years old, whose perambulator had been struck by a runaway truck.

One afternoon some friends of my parents took us for a sight-seeing drive. Somewhere along the way I heard one of them point out a nearby building to my father and remark that it was a jail.

"But why is there a jail here?" I asked. "Isn't everyone Jewish?"

The adults chuckled. It must have embarrassed me because I remember it after almost 60 years. My mental model of Jews, formed by conversations at home, didn't contain scenarios in which we committed crimes.

———

In my 1950s childhood South African Jewish adults were mostly immigrants, a zeroth generation with heavy, embarrassing foreign accents. They had begun their new lives in the poorer mixed-race suburbs and worked hard in small businesses. My father, who arrived in 1934, soon began running Union Service Station, a garage that sold petrol, oil, and batteries, as well as secondhand axle-and-wheel sets for the donkey carts that many peddlers still used. He was ambitious and inventive. During the Second World War there was a shortage of imported car batteries in South Africa, and so he set about learning how to manufacture batteries in a room behind his garage. He obtained molds, melted down solid lead, and cast his own thin flat plates, then immersed them inside black Bakelite battery casings containing a solution of dilute sulfuric acid. These he sold under his own brand with his own warranty. I recall the plates clearly, each a silvery grille you could see through, the glossy lead perforated so as to increase the surface area in contact with the acid. In those days batteries were unsealed, and the garage attendants who filled your gas tank would unscrew the battery tops, check the acid concentration with a glass hydrometer, and then top it up as necessary. I have a clear picture of my father's white lab coat riddled with the brown-edged holes of acid burns. Later, when I studied chemistry in high school, he told me that the correct method of dilution was to pour concentrated acid into water rather than water into concentrated acid, a water splash-back being infinitely preferable to an acid one.

Some of my parents' friends had been in concentration camps and bore the proof of it on their arms. The wife of my bar mitzvah teacher had had her tattoo surgically removed, and you could see the skin discoloration that resulted. Her husband kept his number. Most people I knew had lost close relatives in the Holocaust. Just about everyone was a Zionist, and almost all of these people had relatives who had emigrated to Palestine from Europe. I remember what must have been the 1948 Cape Town celebrations accompanying the establishment of the state of Israel. My Israeli cousin who lived with us for a year lifted me up onto a festive float that was part of an Independence Day parade at the Rosebank fairgrounds. I can still feel her hands in my armpits as she raised me.

I grew up in what amounted to a voluntary Jewish ghetto. Traditionally, Jewish kids in the Diaspora attended daily secular schools and then, several times a week, went to a cheder for late-afternoon Hebrew and Jewish studies. My parents sent me instead to the recently founded Herzlia Day School, a full-time school that combined both a secular and a Jewish education under the same roof. The school was named after Theodor Herzl, the worldly Viennese Jewish journalist who organized the first Zionist Congress in Basel and proposed the creation of a Jewish state 50 years before it finally came into existence in 1948. Our school's motto was from Herzl: "If you will it, it is no legend." In addition to learning Jewish history and reading parts of the Bible in classic Hebrew, we learned to speak, read, and write modern Hebrew, expertly taught by a series of visiting teachers from Israel who rotated through South Africa for a few years at a time.

Though most of our parents adopted the Zionist model, their Zionism came in various political flavors. My parents and many of their friends belonged to Poalei Zion (Workers of Zion), also called the Zionist Socialist Party, which supported David Ben-Gurion and his political Labor movement in Israel in the 1950s. Parents of other friends were Revisionists, so named by Ze'ev Jabotinsky in the 1920s, when he invented his own brand of right-wing Zionism. The Revisionists' slogan, which I heard often, was "A Jewish state with a Jewish majority on both sides of the Jordan," a view that seemed pointless and funny to me in the 1950s and early 1960s, but became much less so after the Six Day War of 1967. The Revisionists were affiliated with the right-wing Herut (Freedom) Party in Israel, led by Menachem Begin, who, before Israeli independence, had led Jewish terrorists against the British colonizers of Palestine. According to my mother, Begin had dated her sister, one of my Israeli aunts, back in Poland when they were both young.

"Socialist" taken seriously would have been a loaded adjective in apartheid-era South Africa. The Cape Town Zionist Socialists were not really Socialist at all; they were not putting themselves on the front line for justice and equality in South Africa. They were petit bourgeois businessmen and their wives, political and intellectual descendants of the prewar *echt* European Zionist Socialists.[6] They held evening teas or fund-raisers once a month in someone's living room, where they all addressed each other as *Chaver* (Comrade). Sitting upstairs in my bedroom while they held a meeting in our living room, my teenage friends and I chuckled condescendingly to hear them call my businessman father "Chaver Derman." We referred to the whole bunch of them half-affectionately, half-mockingly as the *chaverim*.

———

But from age eight to nineteen or twenty I was a junior *chaver* myself. I belonged to Habonim (the Builders), a coeducational Zionist youth movement. Habonim was Lord Baden-Powell's colonial Boy Scouts with the Mowgli mythology replaced by an evangelical pioneering leftish political Zionism, overlaid with the back-to-nature romanticism of the German *Wandervögel* movement of the early twentieth century. The organization was founded in 1929 in England, whence it spread rapidly around the world. We called it "the movement," and it now seems remarkable to me that we let so politically ambitious a phrase fall so easily from our lips.

The movement's aim was that its members fulfill *chalutzik aliyah*. The Hebrew word *aliyah* means "ascension," a metaphorical expression for going to live in Israel, a spiritually higher place. *Aliyah* is also the religious term for ascending to the *bimah,* the platform in the center of the synagogue from which one reads directly from the Torah on Saturday morning, a privilege given to seven people each week. *Chalutzik* is a bastardized adjectival form of *chalutz*, a "pioneer." *Chalutzik aliyah* therefore means a pioneering emigration to Israel. Pioneers set out into new territory to prepare the way for others to follow, which is indeed what the early Jewish immigrants from

Europe to Palestine did in the late 1800s. The movement wanted us to do the same: go to Israel and live on a kibbutz in a communal Socialist framework.

Habonim was merely one of five Jewish youth movements in the Diaspora in general, and in South Africa in particular. Similar to Habonim, but more left and therefore smaller, was Bnei Zion (Sons of Zion). The two groups eventually merged. Even more admirably and rigorously left was Hashomer Hatzair (The Youthful Guard), founded in Galicia in 1913, another movement in the communal Socialist mold but much more severe and radical than Habonim. On the right of Habonim was reactionary Betar, its name an acronym for Brit Yosef Trumpeldor (the Covenant of Joseph Trumpeldor). Trumpeldor, we learned at Herzlia High School, was a one-armed Jewish hero who fell fighting the Arabs in the battle of Tel Hai in Palestine in 1920, exclaiming as he died, "It is good to die for one's country." Just as Habonim was the youth movement allied to Ben-Gurion's Israeli Labor Party, so Betar, founded by Ze'ev Jabotinsky in Riga, Latvia, in the 1920s, was the youth wing of Begin's Herut. Orthogonal to the entire left-to-right political spectrum was Bnei Akiva (Sons of Akiva), a Zionist youth movement whose members were religiously observant, named in honor of the Jewish martyr Rabbi Akiva.

Habonim was highly structured and, most impressively, run entirely by boys and girls in their late teens. There must have been several thousand members countrywide, divided into three age groups: eight- to twelve-year-olds were called Shtilim (saplings); thirteen- to sixteen-year-olds belonged to Bonim (builders); and those sixteen and older were called Shomrim (guards) and administered and headed the movement. They organized the business side of it, coordinated weekly group meetings, planned winter and summer camps, arranged educational trips to Israel to work on kibbutzim, held annual youth congresses, and more, with virtually no adult help. The movement held weekly group meetings for kids in each suburb that had enough attendees to support one. Each group was run by an older teenage *madrich* (guide) or *madricha* (the feminine version).

Some more idealistic members would spend a year or two working full-time for the movement, on salary, in our downtown office. "I'm going to Office," someone might remark when he or she went in to do some work, as though there were only one office in the entire universe. Office was also a good place to socialize. We typed articles and manifestos on waxed stencils and printed copies of songbooks, syllabi, and literary magazines on rotary Gestetner machines.

I was deeply involved in Habonim for my entire life in South Africa. As a child I attended Sunday morning meetings of our local Shtilim group, where I learned classic Boy Scout British Empire skills: tying knots, pitching tents, making fires, building camp furniture out of felled saplings lashed together with string and rope, signaling with semaphore flags. We learned Jewish songs and Jewish history and Israeli geography. We attended outdoor camps for three weeks in the summer and indoor seminars in old up-country hotels for ten days in the winter, drinking hot cocoa boiled in a cauldron and singing around the campfire. We were not so subtly indoctrinated with a go-to-Israel-when-you-grow-up theme, a message that became more explicit as I moved into the group of twelve- to sixteen-year-olds. After that, if you still belonged to the movement and hadn't totally succumbed to the obligations of study, the challenge of South African politics, and the attractions of serial dating, you became a member of the highest age group, the Shomrim. That's the route I took.

Just as the Boy Scouts had Mowgli-related archetypes for elements of its framework, so Habonim had its own Hebrew pioneer words for everything official and ideological. The movement's motto was *Aleh U'vneh,* "Go up and build," and the appropriate response was *Aloh Na'aleh,* "We will indeed go up." The first line of the movement's archaic-sounding song was "Habonim, strong builders, we lads have become," the lads being a nice Scottish Jewish touch.[7] I recall a couplet somewhere in the song that went "We pause not for laggards but build, brick by brick, / A mighty foundation with shovel and pick." Being mostly normal lads and lassies despite all of the ideology, we invariably sang the last phrase as "shovel and prick."

Like middle-class adolescents everywhere, in the final years of high school we concentrated on studies, social life, dances, and the opposite sex. We went to birthday parties, invited dates to see Doris Day and Rock Hudson in *Pillow Talk,* took dancing lessons at Arthur Murray to prepare for school dances (as we called our proms), quickstepped to "It Happened in Monterey," and rock-and-rolled to "A Taste of Honey." British-style, we decided at age seventeen what we (thought we) would do for the rest of our lives and then applied to university to do it. Mostly male would-be doctors went directly from high school to medical school, at age eighteen or nineteen dissecting corpses and examining the insides of women. Regular kids after graduating from high school ignored idealism and proceeded to adulthood along conventional routes; the more politically conscious worked against apartheid. My friends and I, though we participated in many activities outside Habonim, remained in the movement.

Our reasons were many. Some small number of us were truly Zionists, intent on going to Israel. An even smaller subset were Zionist and Socialist, intending to live on a kibbutz. A substantial fraction of the rest of us, socially immature and uncomfortable with the complexities of late adolescence, sought, in the warm womb of the movement, sublimation and a respite from the stresses of social life. The benefits were twofold: we gained shelter from dating and from the perilous thrills of sexual experimentation, and we avoided having to take a stand in an unjust South Africa.

The sexual revolution came to white South Africa later than Philip Larkin's annus mirabilis of 1963, and to the members of Habonim perhaps a little later still. I don't mean to say that no one was interested in sex, but Habonim mores were tinged with a left-wing puritanical morality that developed in the 1940s and persisted through the mid-1960s, at which time I finally left South Africa for the United States. There wasn't much one-on-one dating, which was vaguely discouraged; social life was focused on groups, though some couples did form within them.

But somewhere inside us we scorned what we thought of as bourgeois pursuits. We were taught *Wandervögel*ish slogans and principles from the 1930s or earlier. "A member of Habonim is close to nature and simple in his ways" was one of the more memorable ones. There was an unwritten prejudice against makeup for girls; it wasn't natural. We sanctimoniously looked down on normal interests and ambitions. The movement's highest aspiration was to upend the traditionally Jewish social structure of labor, which, we were taught, was an unfortunate inverted triangle, its top disproportionately heavy with professionals and brain workers and its bottom too light with the agricultural and manual laborers that should have provided a stable societal base. There should be more workers and fewer *luftmenschen,* said the *luftmenschen.* Labor was noble. The best thing you could do was emigrate to Israel, live on a kibbutz, and earn your keep by manual labor in a communal setting. Some young men of my generation chose to become fitters and turners or plumbers rather than go to university. For several years the movement ran a *hachsharah* (preparation camp), a communal kibbutz-style farm in South Africa where you could live and learn agricultural skills in order to prepare for kibbutz life in Israel. We debated the merits of bringing up children in a unit separate from their parents, as happened on some kibbutzim. It was all serious, admirable stuff. While we sublimated we debated ideology, and it *was* stimulating.

*Chalutzik aliyah* wasn't as unreasonable as it may sound now, 50 years later; hundreds of Habonim members eventually emigrated to Israel, and many went to live on a kibbutz. We were living shortly after the Germans had exterminated six million Jews who didn't have a homeland. Furthermore, Jews were disproportionately prominent and active as white foes of the Afrikaner government, some of whose leaders had been pro-German during World War II. As a result it wasn't illogical to think about leaving South Africa, a racist country apparently destined to undergo a bloody final act to its drama of white domination. Trying to sidestep the next Holocaust was a logical move, especially if you had escaped the previous one. As for Socialism, it sounded fair and attractive.

For me this cloistered and romantic haven came to a crisis during my final years at university. Since high school my social life had revolved largely around Habonim. In 1962–1963, when I was seventeen, I spent six weeks touring Israel on an educational program, having fun while learning Zionist ideology and working on Kibbutz Yizre'el in the Galilee, where many of the members were South African. I spent winter and summer vacations working so hard and so happily as a *madrich* that I was too pleasantly exhausted to ponder personal problems. The years flew by; weekends involved Friday night discussions among contemporaries, Saturday night folk dancing and parties with our own entertainment and skits, Sunday mornings or evenings running a weekly meeting for a group of younger kids. Late at night we went to drive-in restaurants for toasted cheeses, chips, and milk shakes and sat in cars talking about intellectual stuff, morality, and girls. It was fun.

And yet I wasn't at all sure that I wanted to emigrate to Israel, and I certainly didn't want to give up studying physics in order to live on a kibbutz. The leaders of the movement, though, had no doubt about what was right. They instituted an *aliyah* register, an oath you had to sign in order to continue to be a member of the movement, your signature certifying that you intended to fulfill *chalutzik aliyah* or, failing that, at least some kind of bourgeois *aliyah*. They argued with members who wouldn't sign it, scorned those who didn't agree with them, were willing to shame them. Late one night, as we sat in a car, a male friend of mine couldn't hold back a burst of frustrated tears after being humiliated by their confident judgments.

"Give me a girl at an impressionable age and she is mine for life."[8] I wouldn't be writing about the movement now if it hadn't left its marks on me, many of them good. But I took all the moral issues seriously, and I very much resented being judged. So somewhere around the age of nineteen, a little bitter, I departed the movement, opening up a deep hole in my social life. What bothered me most was the self-righteous, I-know-what-you-should-do attitude of the few people at the head of Habonim. They were scornful of people with different aspirations, accusing them of wrong thinking or hypocrisy; they were certain of the future and the justness of their arguments, sufficiently so to humble anyone who didn't think their way.

––––––

Ten years later I had completed a PhD in the United States and was now a postdoc at Oxford University. I reconnected with some old South African Habonim friends in London. My plumber friend who had indeed gone to a kibbutz had shortly thereafter abandoned both kibbutz and Israel in order to marry a woman who wanted to live in London. The head of the movement had left the kibbutz too and was also in London, working on a PhD in sociology. None of them seemed to have any compunction about having changed their minds.

## A LOOK AHEAD

Though we tend to rely on them, models fail and theories are almost never perfect. This book is therefore about models and theories: their nature, what to expect of them, how to differentiate between them, and how to cope with their inadequacies. Chapter 2, "Metaphors, Models, and Theories," introduces and analyzes two ways of understanding the functioning of the world. As mentioned at the start of the present chapter, models, like metaphors, tell us merely what something is *like*; theories, in contrast, attempt to tell us what something actually *is*.

Chapter 3, "The Absolute," focuses on the nature of theories, which I illustrate by using Baruch Spinoza's analysis of human passions and the pain they bring. The work of Spinoza, a seventeenth-century philosopher, bears a close relationship to geometry and to the twentieth-century theory of financial derivatives.

Chapter 4, "The Sublime," recounts the development of the most accurate theory in physics: the theory of the electromagnetic field. I show that intuition plays a major role in the discovery of nature's truths.

Chapter 5, "The Inadequate," returns to models, in particular the Efficient Market Model of finance, which has been cited as one of the causes of the financial crisis. I analyze the metaphorical nature of the model's assumptions and point to the places where they fall short. Theories can sometimes be perfect, but models are always inadequate, and financial models especially so.

Chapter 6, "Breaking the Cycle," suggests ways to cope with the shortcomings of models. To work around their inevitable flaws requires a clear understanding of their precepts; it also requires common sense and, especially, ethical principles. I have reprinted a part of *The Financial Modelers' Manifesto,* developed several years ago with a colleague, which proposes a set of principles for financial analysts to live by. An Appendix, "Escaping Bondage," provides a short diagrammatic summary of how Spinoza's theory of the emotions leads to his philosophy for escaping the painful confines of the passions.

## TWO IMPOSSIBLE THINGS BEFORE BREAKFAST

The longer you live, the more you become aware of life's contradictions and of the inability of reason to reconcile them. A therapist friend told me that when she treats patients who radiate negative energy she places 10 cc of water in a glass between her and them in order to protect herself. The physical nature of water molecules, she explained, allows them to absorb negative energy. When her patient leaves, she flushes the water down the toilet. It's an appealing notion, albeit an entirely fanciful one, in my opinion. But my friend believes it's genuine physics and recommended the same strategy to me.

I have a hard time being patient when people confuse metaphor with fact. How do you respond to someone who sincerely believes what she says, and who is trying to help you, but who can't see the boundary between reality and fiction?

In Nabokov's *Lolita,* Humbert Humbert violates boundaries much more dangerous than that. Humbert is a self-described vile creature who craves Lolita for her nymphet body and soul. After she's escaped him, five or more years later he tracks her down and discovers that she is no longer a nymphet at all, but a grown-up, practical and matter-of-fact, worried about money, thickened and pregnant with the child of a simple, almost stupid, unglamorous man. Humbert observes of her:

> There she was with her ruined looks and her adult, rope-veined narrow hands and her goose-flesh white arms, and her shallow ears, and her unkempt armpits, there she was, hopelessly worn at seventeen, with that baby . . . and I looked and looked at her, and knew as clearly as I know I am going to die, that I loved her more than anything I had ever seen or imagined on earth, or hoped for anywhere else.

How can one integrate these contradictions? You have to judge Humbert as responsible for his actions, and yet undoubtedly he's been in the grip of forces beyond his control. Somehow these two irreconcilables—his perversity, which he knows is wrong, and his inability to sublimate it—become partially transmuted, at least in literature, by love.

In life there isn't always such an easy resolution. One has to treat people as responsible for their actions, and yet also recognize that they can't help what they do. It's always easier to regard others from the outside. But one can also try to imagine them as they experience themselves, as we all do, from the inside. Then it becomes possible to see that we all deserve mercy. *Tat tvam asi.* Thou art that.

# CHAPTER 2

# METAPHORS, MODELS, AND THEORIES

*Language is a tower of metaphors • The hole in the Dirac sea • Metaphors become real: the discovery of the positron • Absence is a presence • Analytic continuation • Every fact is a theory • Building a model airplane • Why is a model a model? • Why is a theory a theory? • A puzzling case of monocular diplopia • Making the unconscious conscious again*

## THE DIRAC SEA

### The Metaphorical Rests on the Physical

*Sleep is the interest we have to pay on the capital which is called in at death; and the higher the rate of interest and the more regularly it is paid, the further the date of redemption is postponed.*

So wrote Arthur Schopenhauer, comparing life to finance in a universe that must keep its books balanced. At birth you receive a loan, consciousness and light borrowed from the void, leaving a hole in the emptiness. The hole will grow bigger each day. Nightly, by yielding temporarily to the darkness of sleep, you restore some of the emptiness and keep the hole from growing limitlessly. In the end you must pay back the principal, complete the void, and return the life originally lent you.

By focusing on the common periodic nature of sleep and interest payments, Schopenhauer extends the metaphor of borrowing to life itself. Life and consciousness are the principal, death is the final repayment, and sleep is *la petite mort,* the periodic little death that renews.[1] Life is a temporary *non*blackness.

Schopenhauer's metaphor is striking, but less obvious metaphors are everywhere. Most of the words we use to describe our feelings are metaphors. To say you are "elated" is to say you feel *as though* you have been lifted to a high place. "Feeling high" is an out-of-control version of elation. But why is there something good about being elevated? Because in the Earth's gravitational field[2] all nonfloating animals recognize the *physical* struggle necessary to rise, and when you rise you can see the world spread out beneath you. Being elated is feeling as though you have overcome gravity. Conversely, when we feel depressed we feel as though we have been pushed down to a low place. Things are looking up, we say, or looking brighter, or less dark. These are metaphors too, rooted in our physical senses. Some metaphors are nested, traveling through several layers to their base. When we say the economy is depressed we are comparing the economy's *spirits* (another metaphor) to those of a person who feels *as though* he or she were pulled down by gravity.

Language is a tower of metaphors, each "higher" one resting on "lower" ones that preceded it. Not every word can be a metaphor; you cannot sensibly define every word in terms of other words, or else language would be meaningless. At the base of the tower are words like *push* and *down,* two of the nonmetaphorical words and concepts on which the tower rests. *Push* and *down* are understood with our bodies, because we are wetware, an amalgam of chemicals rather than silicon chips and computer code, and we experience the world through the sensations that chemicals are capable of. You cannot have lived without knowing what it is to have struggled against gravity or felt the insecurity of darkness. That is how we know that *down* and *dark* are bad and *up* and *light* are good.

Had life arisen[3] in outer space, free of gravity and light, there would be no perceptible up or down, and hence no depression or elation. You could be disheartened, perhaps, but not depressed. You could feel full or empty but not light or heavy, bright or dark. And you couldn't take a dim view of your surroundings.

### The Discovery of the Positron: Metaphors Become Real

Just as life can be viewed as a hole in the sea of darkness, so, almost a century later, Paul Dirac showed that the positron is a hole in another invisible sea. The Dirac equation, proposed in 1928, was intended to describe the essential nature of electrons in a manner consistent with both Einstein's Special Theory of Relativity of space and time and Erwin Schrödinger's nonrelativistic wave mechanics of matter, the two then recently discovered theories that together described[4] the nature of matter, space, and time. Electrons, the tiny particles that orbit the nucleus of atoms, have negative electric charge and are responsible for all the chemical properties of matter. Dirac's equation represented electrons as fast-moving relativistic objects (thereby getting their space-time properties correct) described by a probability wave (thus matching the quantum nature of their matter). Once he solved his equation, out fell mathematical solutions that miraculously accounted for the previously unexplained fact that electrons had been observed to spin about their own axes. Dirac's equation also explained various small but significant subtleties in the spectrum of light radiated by an excited electron in a hydrogen atom as it emits a quantum of light and drops to a lower energy state.

*But that's not all!* as TV salespeople say. Also emerging from Dirac's equation were solutions that corresponded to electrons that had *negative* energy. Negative energy is what Wall Street would call a deal breaker, because it implies that the world we know is unstable: if an electron is permitted to have negative energy, then any ordinary electron with positive energy is in an excited state relative to one with negative energy, and can therefore emit a quantum of light as it drops down into a negative-energy state. All electrons in the world would therefore cascade downward into states with arbitrarily large negative energy and radiate their way out of visible existence. But the world isn't unstable, so something is amiss here.

———

When computer programmers are confronted with a misbehaving program, they like to argue, tongue in cheek, that it's not an unintended bug but rather a *feature*. To circumvent the instability of his theory, Dirac came up with a bit of jujitsu, an ingenious argument that turned his bug into a feature, his weakness into a strength. He assumed that the void we live in, what physicists call *the vacuum*, is not empty, but is instead filled to the brim with negative-energy electrons, an infinite number of them at all possible negative energies between zero and minus infinity.[5] This jam-packed void, the background against which we live and act, is the metaphorical *Dirac sea*. It's the vacuum, but it's not really empty. It's full of invisible negative-energy electrons, waiting, he realized, to emerge and manifest themselves as soon as someone or something gives them a large enough jolt.

That jolt is a bolt of light. When a photon with enough momentum hits a negative-energy, negatively charged electron in the Dirac sea, it can impart sufficient energy to it so that the struck electron will pop out above the surface of the sea and become visible as a normal electron

| 1 | 0 | 3 | 4 |
|---|---|---|---|
| 6 | 2 | 11 | 10 |
| 5 | 8 | 7 | 9 |
| 14 | 12 | 15 | 13 |

with positive energy. Having emerged, it leaves behind a hole in the sea, like the hole made by an empty square in a magic number puzzle. As you move the numbered squares around, it is the hole, the absent square, that seems to do the moving. Similarly this *absence* of an electron, this hole, moves around in the sea, and it is the hole itself of which we are aware. Just as an empty square behaves like a square, so this absence of a negatively charged electron behaves almost exactly like an electron, except that, by virtue of its absence, it appears to have *positive* charge. It is an *antiparticle.*

Renowned physicists at the time were highly skeptical of Dirac's sea. Then, in 1932, Carl Anderson at CalTech discovered a positively charged particle in cosmic rays that, except for the sign of its charge, behaved exactly like an electron. He wasn't looking for it, but it was Dirac's antiparticle, the *positron,* the jewel in the theoretical crown and the hole in the sea. Dirac received the 1933 Nobel Prize in Physics, sharing it with Schrödinger, and Anderson received his in 1936, sharing it with Victor Hess, who discovered the cosmic radiation that became the source of many soon-to-be-discovered additional particles.

**Absence Is a Presence**

Schopenhauer's notion of eternal sleep as normality and life as brief temporary periods of punctuated antisleep corresponds to Dirac's picture of the positron as a brief fluctuation in the vacuum. Schopenhauer saw the bad things in life—sickness and pain—as positive, in the sense that they are primary. He saw the good things—health and pleasure—as the mere secondary absence of the bad:

Just as we are conscious not of the healthiness of our whole body but only of the little place where the shoe pinches, so we think not of the totality of our successful activities but of some insignificant trifle or other which continues to vex us. On this fact is founded what I have often before drawn attention to: the negativity of well-being and happiness, in antithesis to the positivity of pain.

I therefore know of no greater absurdity than that absurdity which characterizes almost all metaphysical systems: that of explaining evil as something negative. For evil is precisely that which is positive, that which makes itself palpable; and good, on the other hand, i.e. all happiness and all gratification, is that which is negative, the mere abolition of a desire and extinction of a pain.

This is also consistent with the fact that as a rule we find pleasure much less pleasurable, pain much more painful than we expected.

A quick test of the assertion that enjoyment outweighs pain in this world, or that they are at any rate balanced, would be to compare the feelings of an animal engaged in eating another with those of the animal being eaten.

---

Good as the absence of evil suggests a certain boredom that may not be tolerable for long.

Is good the opposite of evil, the absence of evil, or simply independent of evil? Schopenhauer's perception of the primary status of the negative was reflected again a century and a half later by the Israeli poet Yehuda Amichai:

### The Precision of Pain and the Blurriness of Joy

*The precision of pain and the blurriness of joy. I'm thinking*
*how precise people are when they describe their pain in a doctor's*
*office. Even those who haven't learned to read and write are precise:*
*This one's a throbbing pain, and this one's*
*a wrenching pain, and this one gnaws, this one burns and*
*this is a sharp pain and this*

*is a dull one. Right here. Precisely here, yes, yes.*
*Joy blurs everything. I've heard people say*
*after nights of love and feasting, It was great,*
*I was in seventh heaven. And even the space man who floated*
*in outer space, tethered to a space ship, could only say, Great,*
*wonderful, I have no words.*
*The blurriness of joy and the precision of pain—*
*I want to describe with a sharp pain's precision*
*happiness and blurry joy. I learned to speak among the pains.*[6]

Leszek Kolakowski, the Polish philosopher and historian who lived through Stalinism and died in 2009, also regarded evil as a positive quality:

The Devil is part of our experience. Our generation has seen enough of it for the message to be taken extremely seriously. Evil, I contend, is not contingent, it is not the absence, or deformation, or the subversion of virtue (or whatever else we may think of as its opposite), but a stubborn and unredeemable fact.

G. K. Chesterton experienced goodness as more than the mere absence of badness. In his essay "A Piece of Chalk" he wrote:

Virtue is not the absence of vices or the avoidance of moral dangers; virtue is a vivid and separate thing, like pain or a particular smell. Mercy does not mean not being cruel, or sparing people revenge or punishment; it means a plain and positive thing like the sun, which one has either seen or not seen.

Chastity does not mean abstention from sexual wrong; it means something flaming, like Joan of Arc. In a word, God paints in many colours; but He never paints so gorgeously, I had almost said so gaudily, as when He paints in white.

Goethe, who conducted his own experiments on the perception of color, noticed that when white light is split by a prism or a diffraction grating, the colors of the rainbow arise at the boundaries between light and darkness. Just as electric charges can be positive and negative, and as magnetic poles can be north or south, so, according to Goethe, darkness is the polar opposite of light rather than its absence, and colors arise from the interaction between the poles.

As I recounted in chapter 1, the apartheid government saw white as the positive quality and blackness as the lack of it. In modern disagreement, a Broadway musical whose billboard I walked by a few days ago near Times Square advertises *Spider-Man, Turn Off the Dark*. Which of these views is correct? All, probably. Metaphors are analogies, focused on one quality of a phenomenon but not the entire phenomenon itself. Hence in Schopenhauer's analysis of sleep, it is sleep's periodicity that resembles the coupons of a bond. Like adages, metaphors capture only partial truths, not entireties. As schoolboys in love we used to revel in the conflicting adages "Absence makes the heart grow fonder" and "Out of sight, out of mind," recognizing the partial truth of both of them.

Spinoza, as we will see one chapter hence, is more evenhanded. His theory of emotions regards both *pleasure* and *pain* as independent qualities of human experience, neither one being either the reflection or the absence of the other. If Spinoza is correct, it must be possible to experience both pleasure and pain simultaneously rather than as opposites. I think I can.

Abandoned lovers and lapsed believers can testify only too well that absence is indeed a presence.

**The Positron as Metaphor, Fact, and Theory**

Dirac began with an equation, simple and elegant: $-i\hbar\gamma^\mu\partial_\mu\psi + mc\psi = 0$

Making it work correctly required its interpretation as a metaphor: the sea. This combination, theory plus metaphor, successfully predicted the existence of a particle no one had seen before. A metaphor grounded in a theory can have more power than either alone.

 Dirac found the positron to be a hole in the sea of electrons. He could, had he started with positrons, have found the electron to be a hole in a sea of positrons. Either view works. The key notion is that of symmetry, the absence of one requiring the presence of the other.

The later development of quantum field theory, also pioneered by Dirac, treated electrons and positrons more evenhandedly and less picturesquely, describing both of them as the oscillations of a quantum field that extend throughout space and time, and led to the same results as those of the Dirac sea, but with less need for imaginative effort. Theories discovered by great leaps of individual insight eventually become transformed into formulas anyone can learn.

## ANALYTIC CONTINUATION

Schopenhauer viewed sleep as the metaphorical interest on a loan because of their similar regularities. Taking an analogy based on matching regularities and then extending it into distant regions is a time-honored trick of mathematicians. It's called *analytic continuation.*[7]

Physicists love to extend their theories too, and to impose their extended definitions on us without drawing attention to the subtle transformation. We measure household distances with a ruler or tape measure, but how do you measure the distance to faraway galaxies that the science sections of newspapers so merrily quote as being 100 million light years away? You can't lay out measuring sticks across the universe.

The distance to a galaxy is also an unspoken kind of analytic continuation. One of the ways galactic distances are measured is by observing Cepheid variables, stars whose visible brightness varies. Their true luminosities ("luminosity" is the technical term for their light output, or brightness) have been found to pulsate in a predictably regular way, so that the frequency of their pulsation depends on their luminosity. By measuring the frequency, you can tell something about the true luminosity of these stars. I say "true" luminosities because I want to distinguish between true and apparent luminosities. The true luminosity is the actual light emitted by the star; the apparent luminosity is how bright the star looks, as determined by the light that enters your eye. The farther away a star is, the less light from it reaches your eye. Because the light from a star a distance $R$ away radiates out over a sphere of surface $4\pi R^2$, the apparent luminosity decreases with distance inversely proportional to $R^2$. When you look at a Cepheid variable in a distant galaxy through a telescope, you see its apparent luminosity, but the frequency of the pulsation tells you its absolute luminosity. From the ratio of the true and apparent luminosities you can calculate the distance $R$ to the star.[8]

What an indirect way this is of measuring something as apparently simple and intuitive as distance! The distance to a galaxy has been determined by making use of a regularity of these weird stars that links the quantity of light emitted to the frequency of their pulsation, a "law" that is believable because it can be explained by plausible models of stellar evolution. This measurement of distance makes use of advanced physics rather than Pythagorean geometry. Intergalactic light-years, the circumference of the Earth, the gap between my head and the screen on my laptop, and the separation between atoms—each of these distances is "measured" rather than observed by different methods. Most of these measurements involve the analytic continuation of the notion of distance through the use of models and theories.

I like this observation:

> The ultimate goal would be: to grasp that everything in the realm of fact is already theory.
>
> —Goethe, *Maxims and Reflections*

## DIG WE MUST

 Why models? Because the inanimate world is filled with quasi-regularities that hint at deeper causes. We need models to explain what we see and to predict what will occur. We use models for envisioning the future and influencing it.

The world of people is unpredictable and begs for divination as well. At every moment we face choices with uncertain outcomes. Each decision, even one made on the spur of the moment, involves, just beneath the surface, some imagined model for how the future may evolve and how our choices will affect it. We are always weighing the odds, estimating the relative importance of causality and chance. Without time, there is no need for action.

As time passes, possibilities narrow. Because our lifetime is finite, time, choice, risk, and reward are of the essence. Unless you can live in the perpetual present, you need theories and models to exert some control. Theories and models are a kind of magic, and the builders of successful ones, like Dirac, are shamans bridging the visible and invisible worlds.

## A MODEL AIRPLANE: THE ZIPPY

My earliest recollection of models is of the scaled-down airplanes we used to build from model kits in grade school. When I was eight my mother let me take the bus on my own down to Jack Lemkus in St. George's Street in Cape Town and choose a kit to take home. Some kits were too difficult and time-consuming for an eight-year-old's patience and skills, requiring days of careful assembly; others, the simple gliders, were too unsatisfyingly easy to piece together, taking only a few minutes. One had to find a level of challenge that was difficult and yet surmountable.

The only plane I built successfully was a Zippy. The kit contained long thin strips of lightweight balsa wood used to create the frame of the plane. It also included flat sheets of the same wood with pre-printed cross-sectional inserts that prevented the frame from collapsing. (Tropical balsa is so strong, light, and flexible that the De Havilland Mosquito, a genuine full-size World War II British combat aircraft, was partially constructed of this wood.) A block of balsa had to be carved and sanded and then glued into the nose to hold the propeller. You pinned the plan to your mother's bread-kneading board and used dressmakers' pins to force the long balsa strips to curve along the preprinted arcs that defined the struts. Then you cemented them to each other with airplane glue. When the glue dried, you removed the pins and relied on the cement to maintain the curvature of the stressed beams. Then you glued the sides of the frame to the cross-sectional inserts.

The fuselage was translucent tissue paper cemented to the balsa frame, trimmed, then dampened with water to shrink it taut. When it was dry, you lacquered and painted it to make it stiff and realistic. The engine was merely a long rubber band that ran the internal length of the fuselage, from the propeller block at the nose to a hooked pin inserted into the tail. You rotated the propeller many times to wind up the rubber band and then let it loose. The propeller accelerated and spun as the band unwound, and the plane, it you were lucky, took a brief flight of perhaps ten seconds at best. If you were really ambitious about airplane models—I wasn't, though I admired such ambition in some of my friends—you followed every instruction very carefully, especially sanding off any excess glue on the frame before overlaying the tissue so as to leave no imperfections at all.

I assume that somewhere in the universe of actual airplanes there was or had been a Zippy. My model Zippy was smaller and lighter than the putative actual Zippy; it lacked seats, ailerons, and functioning windows and doors; it was made of totally different materials. Why did they call it a model?

## TYPES OF MODELS

*I'm very well acquainted, too, with matters mathematical,*
*I understand equations, both the simple and quadratical,*
*About binomial theorem I'm teeming with a lot o' news*
*With many cheerful facts about the square of the hypotenuse.*

*I'm very good at integral and differential calculus;*
*I know the scientific names of beings animalculous;*
*In short, in matters vegetable, animal, and mineral,*
*I am the very model of a modern Major-General.*

So sang Gilbert and Sullivan's Major General Stanley in *The Pirates of Penzance* of 1879. The mathematical expertise he sings of is remarkably thorough, not only for 130 years ago but for contemporary financial modelers too. Stanley claims to be what we now call a role model or exemplar, a particular specimen that exemplifies the ideal qualities of a class. That's one use of the word *model*. Model airplanes are another. We also refer to the Model T, fashion models, artists' models, a weather model, an economic model, the Black-Scholes Model, the Standard Model. What do we mean when we call something a model?

## The Model T

The Model T is a type of Ford, one of a class of things belonging to the Ford category. The Model T is an instance, not everything a Ford can be.

## Fashion Models

A fashion model displays clothing or cosmetics. What's important about a fashion model is the exterior: looks, physique, aura. The rest is more or less irrelevant, except insofar as auras and exteriors reflect interior qualities. My daughter was once a hand model in a web advertisement. When you're a model, only parts of you are important. A person is the real thing.[9]

## Artists' Models

An artist's model is a proxy for the real thing. A mannequin is a proxy for a proxy, two degrees of separation. The work of art that uses the proxy is its own real thing, complete in its own way and no longer a proxy at all.

## A Weather Model

A computer model of the weather tries to predict the future weather from the weather today. "Weather" is an abstraction for a collection of an indefinite number of qualities and quantities and the way they vary over the short term, among them temperature, pressure, humidity, and wind speed. A weather model specifies the relevant variables and links them through a set of dynamical equations from physics and chemistry that represent the effects of sunshine, clouds, heat, moisture, evaporation, and air and water currents as they propagate through the atmosphere and along the surface of the Earth as it rotates about its axis and about the sun.

A weather model is much more clearly *not* the weather than the Zippy model is not the airplane. The Zippy is instantly recognizable as a representation of the airplane. The weather model is recognizable as a model of the weather only for someone with the right education.

A weather model's equations are a limited and partial representation of a limitlessly complex system. One cannot model the physics, chemistry, and biology of all the chemicals in the atmosphere and their effect on every species on Earth. There is always the danger that one has omitted something ostensibly negligible whose tail effects over long times are crucially important. This is what makes the predictions of global warming the subject of legitimate debate.

**Economic Models**

An economic model aims to do for the economy what the weather model does for the weather. It too embodies a set of equations to represent the interactions of people and financial institutions. But an economy is an even more abstract concept than the weather. Supply, demand, and investors' utility, to name just a few of many possible variables in the model, are much harder to define (let alone quantify) than temperature and pressure. When you model "the economy" and "the market" you are modeling high-level abstractions.

Friedrich Hayek, the Austrian economist who received the 1974 Nobel Memorial Prize in Economics, pointed out that in the physical sciences we know the macroscopic through concrete experience and proceed to the microscopic by abstraction. For example, the first theories of gases dealt with volume, pressure, temperature, and heat, all directly accessible to our senses. Centuries later we understand pressure as the kinetic energy of invisible microscopic atoms. The atoms, though we consider them real, are more abstract than the pressure and temperature that we perceive directly. In economics, Hayek argued, the order of abstraction should be reversed: we know the individual agents and players from concrete personal experience, and the macroscopic "economy" is the abstraction. If the correct way to proceed is from concrete to abstract, he argued, in economics we should begin with agents and proceed to economies and markets rather than vice versa.

The difficulties one encounters in modeling economic abstractions are illustrated by attempts to deal with the notion of market liquidity. Liquidity is the metaphorical quality that makes trading possible; it connotes the easy availability of counterparties to buy something you want to sell or sell something you want to buy, and its disappearance in states of fear causes the great damage that characterized the recent global financial crisis. Everyone thinks he knows what liquidity means, yet no one has yet adequately defined and quantified it.

**The Black-Scholes Model**

Black-Scholes, as it's commonly referred to by financial practitioners, is the most celebrated and widely used model in all of economics. I spent 17 years of my professional life at Goldman Sachs & Co. extending the Black-Scholes option pricing model in a variety of directions.

A stock option is a kind of lottery ticket you can buy whose future payoff depends on the future moves of the stock price, up or down. It provides reward (if you guess the direction of the move correctly) in exchange for risk (the chance that you guess wrong and lose the price of the ticket). The Black-Scholes Model tells you how to estimate the value of an option in terms of the stock price's risk.

Risk versus reward is the overwhelming issue in finance: how much potential future reward does it take to justify the risk of losing your money when you make an investment? Risk connotes the *possibility* of harm, and so financial theory is intimately bound up with the mathematical theory of probability, which originated centuries ago in connection with the attempt to estimate gambling odds. Buying a stock is a symmetrically risky endeavor: if its market price goes up after purchase, you make money; if the price goes down, you are proportionately harmed. A call option is an investment in only the upside of the stock. If the stock price has risen by some amount at expiration, the option will have made you that many dollars, but if the stock price has dropped, you receive no payoff and lose only the price you paid for the option. The Black-Scholes Model tells you what the value of the option is.

An option is a complex conceptual machine. Its value rises when the stock price rises and falls when the stock price falls. Black-Scholes provides a recipe for *manufacturing* a call by borrowing money to buy shares of the stock. The model tells you exactly how many shares to buy initially and then, at every future instant of time and at every future stock price, how much additional stock to buy or sell so that the stock you own will replicate the payoff of the option contract. The value of the option is the total cost of its manufacture, the cost of all the required trading with borrowed money. The Black-Scholes formula explains how the option value—the estimated cost of trading—depends on the stock price, the interest charged for borrowing, and the riskiness of the stock itself.

Just as a weather model makes assumptions about how fluids flow and how heat undergoes convection, just as a soufflé recipe makes assumptions about what happens when you whip egg whites, so the Black-Scholes Model makes assumptions about the riskiness of stock prices, that is, about how stock prices fluctuate. Black-Scholes assumes that stock prices move smoothly but randomly with a definite volatility, a fixed degree of fluctuation. Given the assumptions, you can figure out the net cost of manufacture. That cost is the fair price of the option, *assuming the validity of the model.* Just as my father could figure out what to charge for homemade batteries by estimating the cost of lead, casting, labor, sulfuric acid, and Bakelite, just as a dessert chef can figure out how much to charge for a soufflé based on ingredients, labor, and waste, so Black and Scholes could estimate how much it would cost to manufacture an option.

But there is a crucial difference between the assumptions made by the Black-Scholes Model and the assumptions made by a soufflé recipe. Our knowledge about the behavior of stock markets is much sparser than our knowledge about how egg whites turn fluffy. Fluids and egg protein don't care what people think about them; markets and stock prices do. Like a weather model (but even more so), Black-Scholes is an ingeniously clever mental model of a complex system, an elegant mechanism that, in trying to reflect the actual world in a short description, must reduce its intricacy. That reduction makes the model usable but simultaneously limits its usefulness.

**The Standard Model**

The Standard Model, for which Sheldon Glashow, Abdus Salam, and Steven Weinberg received the 1979 Nobel Prize in Physics, is a unified description of quarks and leptons, the smallest elementary particles, and the forces between them. The description incorporates into one coherent framework James Clerk Maxwell's nineteenth-century theory of electromagnetism, the 1928 Dirac theory of the electron, and Enrico Fermi's 1934 theory of radioactive beta decay, a framework in which all of these apparently disparate forces are merely superficially different aspects of a single, more general force. I spent the first part of my professional life as a theoretical physicist, working on tests of the Standard Model.

The Standard Model is not really a model at all; it is a description, and hence a *theory*. A theory attempts to provide an accurate portrayal of the nature of things, unifying the outward with the inward, not just saving the appearances but identifying their essence. I say "attempts" because a theory can be right or wrong. What makes something a theory is the way it tries to depict and explain. When someone proposes a model, you can ask "Why?" and expect arguments that make the analogy plausible. When someone proposes a theory, "Why?" is less important. A model is the construction of an analogy. A theory is the linking of the outer with the inner.

The process of unifying several previously disparate theories is a bit like confirming the existence of a never-observed bird from a small fragment of its birdsong. From the song fragment you deduce a morphology; from the morphology you predict the entire song. To confirm the existence of the bird you must then find more fragments of the same bird's song, as predicted. If you hear them, you confirm the theory. The bird itself is never seen.

From fragmentary evidence Glashow, Weinberg, and Salam figured out the entire song; one of its predicted disharmonies was small amounts of *parity violation,* the technical name for a phenomenon in which more particles move to the left than to the right. While *up* and *down* are absolute directions in a gravitational force field, directly perceptible by the human body, until the mid-1950s *left* and *right* had seemed to be conventions of speech rather than physical realities. My left is your right, but our ups and downs are the same. Then in the 1950s physicists discovered that radioactive beta decay, a force whose consequences are also perceptible by the body, does distinguish *left* preferentially from *right*. One can absolutely define *left* and *right* by the direction of the asymmetry in the distribution of particles produced in beta decay. That's a fact.

The Standard Model predicted additional, previously unobserved left-right asymmetries in nature. My PhD thesis of 1973 proposed an experiment to detect these asymmetries in high-energy electron-proton scattering, an experiment in which one smashes spinning electrons into stationary protons and then observes the distribution of the electrons as they bounce off the target. I calculated the size of the predicted asymmetry in the Standard Model. An asymmetry of the appropriate size was finally observed, as predicted, at the Stanford Linear Accelerator Center in 1978. The experiment provided the final stamp of approval and converted the standard model into the Standard Model. The results were welcomed as "the long elegiac salute given to the end of an age."[10] The elegy was the full melody of the Standard Model. The age was the period of pell-mell discovery of new subatomic particles, from the electron in 1898 through neutrons, pi mesons, and their siblings, culminating in the discovery of the quarks inside them and the W- and Z-bosons that mediated their interactions. We are now more than 30 years into the age beyond that. Though physicists can invent many new orchestras consistent with the fragmentary music of gravity and cosmology, none of their instruments has yet been discovered.

## THE NATURE OF MODELS

### There Is Always a Gap

My Zippy wasn't the actual airplane itself, though it bore some similarity to the plane. Similarity lies in the eyes of the beholder and creator.[11] My model Zippy was created with the intention of reproducing some small number of important features on a smaller scale. My Zippy looked like an airplane. Its construction—frame, struts, and fabric to create a light yet strong structure—was sound from an engineering point of view and similar in style (though not in size and material) to the real Zippy. And it could (briefly) fly.

The realistic appearance, the structure beneath the skin, and the ability to fly made the Zippy a suitable model for me at age eight. The structure was important, though reproducing it was hard work. At age three or four I would have been happy with a rudimentary wooden airplane that I could have zoomed through the air with my hand while making throaty airplane noises. If I had been a few years older, I would have wanted a combustion engine and radio control. Had I been an aircraft designer, the ability to test the aerodynamic lift and stability would have been critically important. But, however complex, all of these models are limited when compared with the real thing. There is a gap between the model and the object of its focus. The model is not the object, though we may wish it were.

A dreamed-of counterexample is the model created by Pygmalion: a statue of a woman so beautiful that he fell in love with it. This is a not uncommon occurrence in the worlds of finance and nutrition, both of which abound with experts reluctant to abandon their models in the face of evidence of their unreality. Pygmalion was lucky; Aphrodite granted his request to bring the statue to life, he called her Galatea, and they lived happily ever after.

# I. MODELS

## An Analogy, a Caricature, a Fetish

A model is a metaphor of limited applicability, not the thing itself. Calling a computer an electronic brain once cast light on the function of computers; nevertheless a computer is not an electronic brain. Calling the brain a computer is a model too. In tackling the mysterious world via models we do our best to explain the thus far incomprehensible by describing it in terms of the things we already partially comprehend. Models, like metaphors, take the properties of something rich and project them onto something strange.

A good example is the collective model in nuclear physics, for which Aage Bohr (the son of Niels Bohr), Ben Mottelson, and James Rainwater received the Nobel Prize in Physics in 1975. The collective model regards the core of the nucleus as a drop of dense incompressible fluid that interacts with a small number of so-called valence protons and neutrons outside the core. Of course, the core itself really[12] consists of protons and neutrons held together very tightly by their mutual attraction, but if you think of it as a liquid drop that, when excited, can oscillate, vibrate, and rotate, then you can figure out the energy of its collective excitations and their interaction with the protons and neutrons outside the core. With this model that combined a fluid core with an external shell, Bohr, Mottelson, and Rainwater were able to explain the excited states of uranium and other heavy nuclei

The picture of the nucleus as a drop of water is a limited analogy. Regarding the nucleus as a liquid drop is very different from describing the electron using the Dirac equation. The Dirac equation, even if it eventually turns out to be not quite the absolute truth, will still have been an attempt to intuit the essential nature of the electron. The collective model merely compares the nucleus to a drop of water.

A model is a caricature that overemphasizes some features at the expense of others. It focuses on parts rather than the whole. It is a fetish in which the importance of one key part of the object of interest is obsessively exaggerated until it comes to represent the object's quintessence, such as a shoe or corset standing in for a woman. (Is that perhaps why most modelers are male?) But the shoe or corset isn't the woman; it is just the most important part of the woman for this model user. Once you understand that a model isn't the thing but rather an exaggeration of one aspect of the thing, you will be less surprised at its limitations.

### Let Someone Else's Fingers Do the Walking

Thinking for yourself is hard work, and models save mental labor. Like the vacuum cleaner and washing machine that promised to liberate suburban housewives of the 1950s from drudgery, models provide easy and automated ways of letting other people do the thinking for you.

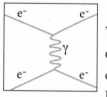

When I worked on my PhD thesis to test the Weinberg-Salam Model in the early 1970s, I carried out each calculation using Feynman diagrams, the cartoonlike representations invented by Richard Feynman in the late 1940s to systematize and enumerate the ways particles interact during collisions. Using a formal set of rules that Feynman developed with his inimitable blend of mathematics and intuition, rules that were later justified by the more rigorous mathematics of Freeman Dyson,[13] I elaborated and drew all the possible diagrams that could occur in the Standard Model, and then, using Feynman's rules, translated each picture into a mathematical formula and evaluated it. The calculations were carried out with pen and paper and took hundreds of pages. To check the accuracy I repeated each calculation at least twice, the second time without looking at the first.

Feynman's diagrams and rules are bookkeeping by picture, a Tinkertoy algorithm that miraculously captures all the details of quantum mechanical forces in the Standard Model via a series of stick-and-vertex diagrams; they allow people less talented than Feynman to use his pictures to perform the most complex calculations carefully and correctly. Like all great advances in physics, they codify and make routine what was formerly almost impossible to think about. Only Feynman could have invented it; now wet-behind-the-ears graduate students can churn out page after page of accurate calculations. Indeed, by the time I stopped doing physics in 1980 I knew two professors in Wisconsin who had programmed computers to generate the diagrams in the Standard Model, translate them into formulas, evaluate them, and mechanically calculate the magnitude of the effects predicted. All that remained was to write the paper.

Einstein similarly made calculations easier for lesser physicists when he discovered the *theory* of special relativity. Hendrik Lorentz, for whom the Lorentz-Fitzgerald relativistic contraction of rapidly moving rods was named, had more or less come to the same conclusions as Einstein. But he did so by carrying out difficult and complex calculations using elaborate *models* of the atoms within rods, determining what happened to the forces between them as they moved. Einstein replaced these struggles with simple but deep analyses of what it means to talk about the length of a rod. Now high school students can calculate the changing size of objects and the changing periods of clocks as they move. Lorentz's model needed justification. Einstein's theory is exact fact.

In both physics and finance the first major struggle is to gain some intuition about how to proceed; the second struggle is to transform that intuition into something more formulaic, a set of rules anyone can follow, rules that no longer require the original insight itself. One person's breakthrough thus becomes everybody's possession.

## A Model Is a Little Language

It takes hard work to master a model. In *Zen in the Art of Archery,* Eugen Herrigel describes his repetitive, initially futile struggles with the Japanese bow and arrow, until finally he was able to transcend the battle for conscious competence and achieve unconscious skill, pulling back the string and launching the arrow mindlessly, carelessly, and accurately at the target. Anyone who plays tennis only occasionally will have noticed that, after a hiatus, one often does better the first time on the court than the second. As one tries to try to make improvements, one gets worse. As Spinoza wrote, "The body can by the sole laws of its nature do many things which the mind wonders at."

The most valuable knowledge is that which has become unconscious and intuitive. Focusing one's eyes, grasping, manipulating, chewing, crawling, walking, or speaking—one begins by struggling to do these things and ends by doing them without thinking or struggling. Practice to the point of automaton-like competence is necessary. Until you can do something without thinking, you can't progress further up the hierarchy of linguistic or modeling metaphors. If you eschew the help of the mental machines or models created by your intellectual forebears, you have to think through everything for yourself, every time. There are occasions when the capacity to think from scratch is important, but most of the time it's best to take your foundation for granted.

So many of our acquired abilities move from conscious struggle to unconscious achievement that some writers have theorized, implausibly but fascinatingly, that every unconscious human ability (even digestion, to take an extreme example) was first learned by conscious efforts of the will, and that it is the failure to achieve unconscious automation that leads to various kinds of mental ailments.[14] Suffice it to say that when you have digested a model or language well—and a model *is*, like language, a framework for communication—then, with it inside you, you gain power.

## Models Reduce the Number of Dimensions

The world is impossible to grasp in its entirety. We can focus on only a small part of its vast confusion. Models project multidimensional reality onto smaller, more manageable spaces where regularities appear and then, in that smaller space, allow us to extrapolate and interpolate from the observed to the unknown. At some point, of course, the extrapolation will break down. What's amazing is how well this strategy of reduction can work, especially in the physical sciences.

Models in finance use the same strategy. Companies that issue stock are multidimensional. You can evaluate them with respect to many metrics: management, earnings, debt, credit quality, patents generated, and so forth. The stock market's job is to collapse all of these qualities and quantities into one number, the stock price, measured on a one-dimensional scale of dollars. But price alone doesn't indicate relative value: Who knows which is a better deal, apples at $2 a pound or oranges at $3? IBM at $100 per share or Microsoft at $250? Financial models attempt to answer these questions. To do so they project the company onto an axis that measures value more usefully than dollars do. Though price is a fact, value depends on the observer. For office space, value might be price per square foot. For fruit it could be taste, or the quantity of vitamins and fiber, depending on who the model serves. For stocks, value can be measured by the ratio of price to earnings; for bonds, by the yield to maturity. All of these axes represent a view on the source of value, and hence a model.

## The Dangers of Extrapolation

Models project a detailed and complex world onto a smaller subspace. But extrapolation in the smaller space can be unreliable. The supposed benefits of estrogen supplements for postmenopausal women and the advantages of margarine over butter for preventing arterial plaque have turned out to be dubious. Estrogen can slow osteoporosis, but it's associated with an increase in various kinds of cancer. Cholesterol intake from food isn't directly equivalent to cholesterol output into plaque, though it is plausible that they are linked. Eggs and butter contain many substances that are good for you, whereas margarine contains many that aren't. Models are simplifications, and simplification can be dangerous.

# THE NATURE OF THEORIES

A weather model's equations are a model, but the Dirac equation is a theory. What *is* a theory, and why do I call the Dirac equation one? The Online Etymological Dictionary lists a definition of the word *theory* dating from 1638 as "an explanation based on observation and reasoning." Wikipedia cites Francis Cornford, an English scholar, suggesting that the practitioners of the Greek religion Orphism used the word *theory* to mean "passionate sympathetic contemplation." Both phrases are well chosen.

Models are analogies; they always describe one thing relative to something else. Models need a defense or an explanation. Theories, in contrast, are the real thing. They need confirmation rather than explanation. A theory describes an essence. A successful theory can become a fact.

The ultimate goal would be: to grasp that everything in the realm of fact is already theory. . . . Let us not seek for something behind the phenomena—they themselves are the theory.

—Goethe, *Maxims and Reflections*

---

In *Science and Reflection,* Martin Heidegger comments on the contemplative nature of theoretical insight:

The word "theory" stems from the Greek verb *theorein*. The noun belonging to it is *theoria*. Peculiar to these words is a lofty and mysterious meaning. The verb *theorein* grew out of the coalescing of two root words, *thea* and *horao*. *Thea* (cf. theater) is the outward look, the aspect, in which something shows itself. . . . The second root word in *theorein, horao,* means: to look at something attentively, to look it over, to view it closely. Thus it follows that *theorein* is *thean horan,* to look attentively on the outward appearance wherein what presences becomes visible and, through such sight—seeing—to linger with it.

The related Greek word *aletheia* means "the state of not being hidden," and suggests that theater (*thea*) hides, and that the role of theory is to make evident what is hidden. That's the way I have felt whenever I've done research, in physics or finance. The creator of a theory is attempting to discover the invisible principles that hide behind the appearances. Evolution is a theory; so are the theory of dreams, Newton's laws of motion, and the Standard Model. A theory is (potentially) deep; a model, even when efficacious, is shallower. Theories describe absolutes, like Moses descending from the mountain with the Ten Commandments, or like God commanding, "Let there be light!"

You can see why intuition, the union between object and subject, bears such a close relationship to theories. A theory doesn't simplify. It observes the world and tries to describe the principles by which the world operates. A theory can be right or wrong, but it is characterized by its intent: the discovery of essence. A theory is an absolute nonmetaphorical insight, and this is why the abstractions of mathematics are often more suitable than words for formulating theories.[15] When I wrote papers in physics, I dreamed of discovering a theory that would be true and transcendental and would survive my lifetime. Newton, Lagrange, Hamilton, Darwin, Maxwell, Freud, Einstein, Bohr, Schrödinger, and Gell-Mann, to name just a few, accomplished this. Therefore it's possible.

Newton's Universal Law of Gravitation is a theory. It postulates that the force between any two masses (the moon and the Earth, the Earth and an apple) is proportional to the product of their masses and inversely proportional to the square of the distance between their centers. Newton's Second Law of Motion, a theory too, dictates that the gravitational force accelerates each particle inversely proportional to its mass. Solving these equations, Newton deduced that planets must traverse elliptical paths around the sun, thereby justifying Johannes Kepler's empirical laws of planetary motion.

Newton's theory is general and precise. The gravitational force is inversely proportional to *exactly* the square of the distance between the planets; Newton was confident that the power of the distance is precisely 2. Had he been a social scientist performing statistical regressions in psychology, economics, or finance, he would probably have proposed a power of $2.05 \pm 0.31$.

A theory is not a fetish; when it is successful (see the quantum theory of electricity and magnetism in chapter 4) it describes the object of its focus so accurately that *the theory becomes virtually indistinguishable from the object itself*. Maxwell's equations *are* electricity and magnetism; the Dirac equation *is* the electron; the Weinberg-Salam model of weak and electromagnetic interactions matches the electrons and quarks in almost every detail, as closely as one can measure. You can layer metaphors on top of the equation, but the equation is the essence.[16]

A theory doesn't have to be complete or unmodifiable. Now, more than 106 years after Einstein's Special Theory of Relativity and 96 years after his General Theory, we know that Newton's laws are not quite accurate. But there is still something recognizably absolute about their intent, and they have not been relegated to the status of a model despite their limitations. They are theories that are not exactly right, but they are not models.

Models splinter when you look at them closely. Theories are irreducible, the foundations on which new metaphors can be built. Theories are the thing itself; when you look closely, there isn't anything more to see. The surface and the object, the outside and the inside, are one.[17]

## MONOCULAR DIPLOPIA

Twenty-five years ago I accepted the offer of a free group tennis lesson while on vacation at a Caribbean resort. The coach put six of us onto a single court and made us practice volleying in pairs with our partners across the net. I was in the pair on the extreme right. The gentleman diagonally across the net from me, on the extreme left, decided to be a hero and slammed the ball at his partner as hard as he could. Instead it flew diagonally across the net and directly into my right eye. I fell to the ground and tried to figure out what was happening.

It hurt, but not too badly, and in the end I ignored it and assumed that the black floaters I saw were just what you should expect to see if you got hit in the eye. A week later, when I returned to New York, I realized that things around me were beginning to look as though I were living in an aquarium. I went to an eye doctor and, just in time, discovered that my retina had a "horseshoe" tear and was beginning to detach, like Scotch tape starting to peel from one corner. An eye surgeon resealed the retina to the back of my eye by welding it with lasers beamed through the front and with liquid nitrogen applied to the back. I still go for regular checkups to make sure my retina stays in place and have been warned to watch out for sudden changes in vision.

One morning 23 years later, in early 2008, I woke up and took a cab to work. Looking at the meter, I realized something wasn't quite in order. Used to testing my vision, I shut my left eye. Everything looked okay. Then I shut my right eye and immediately realized that all the LED digits on the taxi meter were doubled. Uh-oh, I thought. My retina again. I told the taxi to turn around and take me directly to my retinal specialist.

A careful man, the specialist tested my vision, measured the pressure inside my eye, dilated my pupils, visually examined my retina from all angles, had a technician do an ultrasound scan of my retina and the layers below it, and finally had another technician inject a fluorescent green dye into my bloodstream and then film the back of my retina to check the state of its capillaries. He saw nothing awry.

Monocular diplopia, the fancy name for seeing double in one eye, is a rare phenomenon. It cannot be a problem of accommodation or so-called squintness that occurs owing to a muscular imbalance between both eyes. My doctor sent me to an ophthalmic neurologist, who, after carefully checking that I could walk in a straight line, stand on one leg, and follow his index finger with my eyes, concluded that I had no neurological problems and told me to go away and not come back.

Nevertheless the diplopia persisted, waxing and waning inexplicably, sometimes almost disappearing for a few hours and then returning with full force. I had trouble reading. I tested my vision constantly, closing one eye and then the other while trying to read distant street signs. I used my thumb and index finger to create a small pinhole through which I peered. When I did so the signs became sharp, which told me that my retina was working well and that the trouble was therefore with the diffraction. Light from each letter in a distant sign reaches your retina along many parallel paths through your lens and the vitreous fluid, and somehow the paths weren't converging to one retinal point, hence the doubling. My retinologist could see nothing especially wrong with my retina and had no further suggestions for treatment or diagnosis, and so, perforce, I became my own general contractor.

I subcontracted parts of myself out to different specialists, trying to figure out what was happening. I read the package insert on a medicine I had been taking and saw that it listed double vision as a rare side effect. Eventually I learned that almost all drugs list double vision as a possible side effect, but when it occurs it's usually binocular diplopia. A very good optometrist I knew for years found that my astigmatism changed even during the course of a brief visit in which he tested my vision. He told me it might be due to stress. But neither of us could understand how stress could affect the curvature of my lens.

Years earlier I had had a cataract removed from my left eye; its lens was replaced by a plastic one that worked fine. But, I reasoned, a plastic disk floating inside the capsule behind the cornea might have suddenly shifted or tilted a small amount the day the double vision began, and so caused my problem. It might still be loose and moving about each day, hence the fluctuations.

I visited doctors recursively, one doctor suggesting the next. The $n$th doctor was the cataract surgeon who had originally inserted my artificial lens, and I asked him to check whether it had shifted. It hadn't. But he could theorize too: most likely, he said, there were some small, undetectable distortions in the rods and cones of my retina, deformations too small to show up on any retinal scan. The rods, which should have pointed straight up, were most likely askew, blades of plastic grass on trampled Astroturf. Nothing anyone can do about it, he said.

Once he had delivered this diagnosis, something impelled him to measure the curvature of my cornea, a test no one had done in the preceding months. A small optical device shines perfectly circular concentric rings of light onto your cornea and photographs the reflections; their deviation from perfect circles indicates the contours of the cornea. Mine was badly buckled on the nasal side. He informed me curtly and concisely that my problem was not the retina. I had keratoconus, a disease in which the cornea progressively thins and ultimately bulges under its own weight, tugged at by the Earth beneath it. (In that case, I reasoned, I should see better in outer space or in free fall, a diagnostic test I must remember to recommend when space flight becomes cheaper.) The resultant bulges in the cornea refract light improperly and cause the diplopia. I didn't like the name keratoconus, which reminded me of calluses and rough heels, and much preferred the fluid sound of monocular diplopia. Then he washed his hands of me and told me to see a corneal specialist.

Before being examined by the corneal doctor I was prepped by his technician, who took my medical history and carried out preliminary examinations. Agitated by my long search for a diagnosis, I told him about my many months with monocular diplopia.

"Anyone ever look under your eyelid?" he asked me.

"No, never," I said.

Another technician repeated the corneal topography on both my eyes. Then I went into the examining room of the corneal man, who reconfirmed the anomalous bulges and the diagnosis of keratoconus, though he admitted that it was a little peculiar that the bulge was on the nasal side of the cornea: gravity would have demanded a bulge at the bottom. The short-term solution was to wear a hard glass contact lens. Tears would fill the gap between the outside of my distorted cornea and the inside of the rigid glass lens, thereby restoring a perfectly spherical surface for light to enter. If it worked, it would also confirm the diagnosis. The longer-term solution, since the cornea would progressively thin and weaken, was either a corneal transplant or the implanting of a scaffolding of plastic circular struts into the cornea to support it.

Before I left I told my corneal specialist what his technician had said about looking under my left eyelid. He shrugged. A week later I went back to his office to be fitted for contacts. Agitated as usual, I told a different technician my whole story as she prepared to try a hard contact on me to see if it improved my vision. I also told her what the technician had said.

"I can look under your eyelid," she said. She had me lean back and she rolled up my eyelid like a window blind.

"There's a chalazion there," she said. That's a fancy word for a tiny hard lump on the inside of the upper eyelid, a sort of chronic stye. I didn't like the word *chalazion* either; it reminded me of both a venereal disease and a Spanish pork sausage. "It pushes on the cornea every time you blink or close your eyes, like someone sticking a finger into a balloon, and distorts your vision," she explained.

Retina, brain, lens, keratoconus. Everyone I consulted had tried, within and sometimes outside the confines of his specialty, to find some explanation for my symptoms, ignoring the obvious. It took a technician with no preconceptions to offer a commonsense explanation. My specialists were quite unabashed at their misdiagnoses. Because I had had a retinal problem in the past, everyone, myself included, refused to imagine any other cause, especially a simple, mechanical, and easily observable cause, like pressure on the cornea from a bump on the inner eyelid.

I tried the conservative treatment for the chalazion first: hot compresses and antibiotic creams to soften up and shrink the bump. But it had been around too long, and so a few weeks later another specialist cut it out and I've been okay since. Which leads me to the following section.

## MAKING THE UNCONSCIOUS CONSCIOUS AGAIN

We cannot be forever examining our foundations; we look particularly to those places where it is reported to us that they are insecure.

—A. S. Eddington, *The Mathematical Theory of Relativity*

 We become expert at the models and theories we use, unconscious of how we use them. This unconscious grasp then serves as a framework for further conscious advancement. One progresses from grasping an object to pitching a baseball to juggling, each accomplishment starting out as a conscious attempt and ending up as unconscious mastery. Conscious and unconscious play leapfrog, the conscious vaulting over the unconscious to then serve as a new unconscious foundation itself.

Faced with crises, the unconscious must become conscious again. When models produce paradoxes or conflicts, it becomes necessary to expose the taken-for-granted assumptions, *reculer pour mieux sauter*. That's what my clever eye technician did. When I later asked one of the doctors I had visited why he had never thought of looking for a bump on my eyelid, he told me that he had assumed that a bump large enough to dent my cornea every time I blinked would have been visible from the outside. It wasn't true. Only the technician was able to hear the facts as I explained them.

Einstein exposed the unconscious when he examined the foundations of classical physics. Physicists before him had unthinkingly assumed that simultaneity was a self-evident notion: to say that two events happened at the same time seemed a simple thing. Einstein made explicit how one would confirm that two events happen at the same time. This reexamination of fundamentals led him to the Theory of Special Relativity, whereby he discovered that, if the speed of light is the same for all observers, then length and time themselves must vary from one observer to another. When you say it like that, it sounds obvious: How can the speed of light possibly be the same for everyone no matter what speed each observer is traveling at, unless length and time themselves change with motion? It wasn't obvious before Einstein.

When the unconscious assumptions of everyday living begin to conflict with each other, it's time to bring them to the surface. Psychoanalysis aims to make the unconscious visible by talking and introspection. Tibetan Buddhists try to achieve the same result by observing the thoughts bubbling out of the mind. Dropping back is sometimes a good idea. The revision of fundamentals often marks great leaps forward.

## ADDENDUM: GOETHE ON SYMBOLISM

I recently came across some remarks by Goethe in an essay on symbolism that reflect on the limitations of words and metaphors in approaching what he calls nature's inner relationships:

Neither things nor ourselves find full expression in our words.

Something like a new world is created through language, one consisting of the essential and the incidental.

*Verba valent sicut numi.*[18] But there are different sorts of money: gold, silver, and copper coins, or paper money. The coins are real to a degree; the paper money is only convention.

We get by in life with our everyday language, for we describe only superficial relationships. The instant we speak of deeper relationships, another language springs up: poetic language.

In speaking of nature's inner relationships, we need many modes of description. I will mention four here:

## Symbols

1. which are *physically*[19] and really identical with the object: e.g., we have learned to express magnetic effects, and now apply this terminology to related phenomena;

2. which are *aesthetically*[19] and really identical with the object. All good metaphors belong in this category, but we must guard against a display of wit which seems to relate the unrelated instead of finding true relationships;

3. which express a connection which is somewhat arbitrary rather than fully intrinsic; such a symbol, however, points to an inner relationship between phenomena. I would say these symbols are mnemonic in a higher sense, for ordinary mnemonics uses wholly arbitrary notation;

4. which are derived from mathematics. Because they are founded on intuitive perceptions, they can become identical with the phenomenon in the highest sense of the word.

We find instances of the first three symbols in language:

1. when, for example, the word expresses a sound (like the noun bang).

2. when the sound expresses an identical feeling (this often happens in inflected forms: banging).

3. when related words have a similar sound (like mine and thine); such words might be dissimilar (I and thou), but moi and toi are related in this way.

The fourth type, based on intuitive perceptions alone, cannot occur in language.

# II. MODELS BEHAVING

# CHAPTER 3

# THE ABSOLUTE

*The Tetragrammaton • The name of the name of the name • The irreducible nonmetaphor • Spinoza's theory of the emotions • Fiat money • How to live in the realm of the passions*

Theories aim to describe essences and must therefore deal with absolutes, the traditional province of religion, philosophy, and mathematics. The best and most accepted absolutes are the theories of inanimate matter—Newtonian mechanics, electromagnetic theory, relativity, and quantum mechanics—that require mathematics for their exposition. Indeed the development of much of mathematics has been triggered by the needs of physicists elaborating their theories. But although mathematics is the most value-free way to formulate it, a theory doesn't have to be mathematical, and it doesn't have to confine itself to the inanimate. A theory, as long as it makes no analogies, can deal with anything.

This chapter therefore deals with the idea of God, who represents the ultimate ground beneath all metaphors, the literally incomparable. In particular I present Spinoza's profound analysis of the structure of human emotions, the agonies they cause us, and their link to the nature of Being and God. Spinoza's description is absolute, not relative; it stands on its own two feet, and is a perfect and accessible emodiment of what I call a theory.

## THE TETRAGRAMMATON

מוֹרֶה אֲנִי לְפָנֶיךָ. מֶלֶךְ חַי וְקַיָּם. שֶׁהֶחֱזַרְתָּ בִּי נִשְׁמָתִי בְּחֶמְלָה. רַבָּה אֱמוּנָתֶךָ,

Modeh ani lefanecha melech chai vekayam shehechezarta bee nishmati bechemlah— rabah emunatechah.

At an early age I was taught to pray in Hebrew, though no one took the trouble to tell me what the prayers meant. In kindergarten we mindlessly learned to chant the *Modeh,* a children's prayer to be recited on waking that, I now understand, thanks God for restoring your soul after its disappearance during the night:

Translated it reads:

*I give thanks before you, living and eternal king, that you have returned within me my soul with compassion; how abundant is your faithfulness.*

Schopenhauer regarded the void as positive and life as its interim negation. Not so, according to the *Modeh,* which views each person's soul or spirit, *nishmati* in the transliteration above, as the positive substance that God returns to your zombie body each morning. *Neshamah,* the Hebrew word for "soul," is also the word for "breath."

For Schopenhauer, life ends when God asks for His money back, perhaps to recapitalize someone else. As long as He doesn't come knocking at the door to repossess it, life continues. For the *Modeh,* in contrast, sleep is soulless and life doesn't continue unless God acts to reinstill it each morning.

Who is this Jewish God of action, and what are His qualities? Note that nowhere in the short *Modeh* does the Hebrew word for God actually appear. The prayer refers to Him only metaphorically, as the living and eternal king.

But God does have a name, and His name has a name too. The

Tetragrammaton[1] is the formidable and fancy *name of the name* of God. Glorious with gravitas, the Tetragrammaton is the majestic way of referring indirectly to God's name as it appears in the Bible, as a sequence of four Hebrew consonants, יהוה, read from the right to the left.[2] Sequentially transliterated, the four consonants correspond approximately to the English letters YHVH, as follows: יהוה. The first letter, *Y* in English, corresponds to the Hebrew letter yod, י. Written left to right, as they would be in English, the letters form the word YHVH, pronounced *Yahweh*. A single name, with no patronymic, as in Prince and Madonna, is grand. A single name with its own name is even grander. YHVH is a bit like HRM, a Triagrammaton for the king or queen of England.

If you are familiar with Hebrew you will recognize the sense of awe that accompanies a glimpse of the Hebrew word יהוה. It is the letters alone that produce the awe, not their pronunciation, as I will shortly explain. To approximate their aura, I write the English letters of God's name in white on black: YHVH.

Throughout the 12 years in school in which I learned to read and speak Hebrew and to recite prayers by heart, I saw the word יהוה printed innumerable times in Bibles and prayer books. Yet I never once heard the word pronounced *Yahweh*. Everyone *I* knew pronounced YHVH as though it were the entirely different Hebrew word אדני /ADNY, pronounced *Adonai*, as though the letters YHVH were in reality the letters ADNY. The words ADNY and YHVH barely resemble each other. It was as though I was taught to say "HRM" whenever I saw the word "Elizabeth."

I have written the letters ADNY in a workmanlike sans serif typeface to contrast their mundanity with the extraordinariness of YHVH. There is nothing fancy or glorious about the word ADNY. It's simply the possessive plural of *Adon*, the Hebrew word for "lord," or "master," and is therefore roughly equivalent to the respectful "Milord." In the British-style high school I attended we addressed our English teacher as "Sir" and our Hebrew teacher as "Adon," equivalent terms that acknowledge authority but lack awe.

ADNY is one of God's many aliases in the book of Genesis. The Bible sometimes also refers to God as ADNY ADNM, pronounced *Adonai Adonim,* a vocative "My Lords of Lords," more wondrous than *Adonai* alone by virtue of its recursion. Surprisingly, God's aliases in the Bible precede the appearance of His real name. God as *Yahweh* is absent throughout the first chapter of Genesis, whose first line reads "In the beginning *Elohim* created the skies and the earth." *Elohim* is the plural of *El,* a generic god, and means simply gods in the sense of a royal pluralized divinity.

The Tetragrammaton YHVH makes its first appearance in Genesis 2:4, only after God has perfected the Creation and can finally rest:[3] "These are the generations of the heavens and of the earth when they were created, in the day that YHVH *Elohim* made earth and the skies." Even here He carries a double moniker.

To this day when I see the letters YHVH I hear in my head the word ADNY. So deep and thorough is this conflation that not until I was far into adulthood did I realize that the letters YHVH did not literally spell ADNY. I looked at the Hebrew letters for YHVH and my mouth and mind said ADNY. My classmates too, I discovered, suffered from the same confusion, even to this day. We saw one word and uttered the other. No one ever mentioned that the actual pronunciation of YHVH was *not* ADNY.[4]

No observant Jew ever utters the name *Yahweh.* By custom rather than commandment, God's name is not pronounced aloud or even under one's breath. Being brought up customarily, my friends and I were never told not to say His name. We didn't know He had one.

## THE NAME OF THE NAME OF THE NAME

A few of my more observant school friends went a step further: not only did they avoid uttering the name **YHVH**, but they also avoided uttering the alias ADNY. Instead they pronounced *Adonai* as *AdoShem*. The reason was as follows. The ** י**/yod, the last (leftmost) Hebrew letter of **אדני**, is also the first (rightmost) letter of the word **יהוה**. Thus the alias ADNY incorporates one letter of God's true name, too close for comfort. Therefore some Jews replace the ** י**/yod in ADNY with the entire Hebrew word *Shem* (meaning "name"), pronouncing ADNY as *ADoShem*. It's as though the normal honorific HRM used to refer to the queen becomes uncomfortably familiar when it refers to Queen Mary, because of the shared letter *M*, so that, just for Queen Mary, the alias HRM is replaced by HRName.

Orthodox Jews carry this to extremes and avoid pronouncing anything reminiscent of the Tetragrammaton. Thus even the word *Elohim* is pronounced *Elokim,* so that the shared *H* of **YHVH** and *EloHim* is replaced by the cognate but symbolically insignificant *K.*

Many go numerologically further. In ancient Hebrew one uses the letters of the alphabet to represent numbers. The tenth letter of the alphabet is **י**/*Y* and the fifth letter is **ה**/*H,* so that the number 15 is represented in base 10 by the letters *YH.* But *Y* and *H* are the first two letters of God's name, and so no one writes *YH* for 15. Instead 15 is written as *TV,* because the numbers corresponding to the ninth letter, *T,* and the sixth letter, *V,* also add to 15 but don't seem to reference God. (It seems to me that replacing *YH* with *TV* makes the combination TV even more reminiscent of God, the unignorable avoidance indicating a presence. One could go one step further and replace 9 + 6 with 8 + 7, but that would cause the same problem.) Similarly the number 16, 10 + 6, which should be written in base 10 as the Hebrew letters *YV,* shares two letters with **YHVH** and is therefore written *TZ*, 9 + 7. Some don't know where to stop, and write G-d rather than God, as though "God" were God's name. It isn't.

## THE IRREDUCIBLE NONMETAPHOR

Moses, tending the flock of his father-in-law, Jethro, near the mountain of Horeb, saw a burning bush whose flame could not consume it. God, from within the bush, declared Himself to Moses and commanded him to deliver the Israelites from Pharaoh.

"Who shall I tell them sent me?" asks Moses.

"Tell them: I am that which I am," answers YHVH.

*Yahweh* is the antimetaphor, the ultimate ground, for whom no analogies are possible and no similes adequate. He is what He is. *Yahweh* is the name of something that isn't a model of reality, but reality itself.

The excommunicated Jewish philosopher Baruch Spinoza, who was interested in the nature of reality, defined wonder as "the conception of anything, wherein the mind comes to a stand, because the particular concept in question has no connection with other concepts." *Yahweh* has no connection with other concepts; He has no qualities; He is *beyond* intelligence and beauty; He is *beyond* good and evil; He cannot be categorized. He is precisely what He is, and He tells us so.

When He says, "I am that which I am," sometimes translated as "I will be that which I will be," God is riffing on His true name: the Hebrew for "I will be" is אהיה/*EHYH*. Its root is *HVH*, the last three letters of God's name. *HVH* means "being" and is also the name of the present tense in Hebrew grammar. Presence and Being necessarily precede everything. YHVH is the irreducible substance out of which everything else is constructed. "*Yah Weh*" is the sound identical with the object, an inhalation *Yah* followed by an exhalation *Weh*, the sound of panting while running, the vocalization of a woman in bed. You can't ask "Why?" about **YHVH**; you can only attest to His existence.

Like God, the electron is precisely what it is and equally wondrous. About it too you can ask only "What?" not "Why?" For the electron, the wonder is the Dirac wave function $\psi$, the so-called four-component relativistic spinor that satisfies the Dirac equation. The spinor $\psi$ is the alter ego of the physical electron, the idea that corresponds to its material states. Two of its four components describe the state of its internal magnetization (north or south), and the remaining two describe the positron—the electron's antiparticle—and its two magnetizations.


## A THEORY OF THE EMOTIONS

I first began to read Schopenhauer in 1979. I had been willfully ambitious with regard to physics and was just beginning to realize that my passion for it was depriving me of air. Still ambitious, but aware now of the repercussions, I found solace in Schopenhauer's description of the miseries the Will inflicts on all of us. That led me to Spinoza's *Ethics,* published posthumously in 1677, and his analysis of human behavior.

In part 3 of *Ethics,* entitled "Of the Affects," Spinoza tried to do for human emotions what Euclid did for geometry. In the rest of this chapter I want to cast light on Spinoza's theory in order, ultimately, to contrast theories with models. Because all of us experience feelings, exploring a theory of them seems like a good place to start.

Euclid used primitives, ingredients everyone is familiar with, as his raw material. Geometric primitives include points and lines, which Euclid defined as follows:

A point is that which has no part.
A line is a breadthless length.

The extremities of lines are points.

A straight line lies equally with respect to the points on itself.

The definitions are ingenious but obscure. If you didn't know what points and lines were by having had them shown to you, these definitions wouldn't help you envision them. We are embodied and begin with our senses.

To his primitives Euclid added axioms, the apparently self-evident logical principles that no one would argue with, stating, for example, "If equals are added to equals, then the wholes are equal." Since 3 = 3 and 2 = 2, then 3 + 2 = 3 + 2. This is indisputable, but not all axioms are as unquestionable as they appear.

Finally, he proceeded to theorems, the interesting and often unexpected deductions he could prove by applying the axioms to the primitives to construct a chain of reasoning. The most famous is Pythagoras's theorem that relates triangles to squares: the sum of the squares of the two perpendicular sides of a right-angled triangle is equal to the square of its hypotenuse. This method, beginning with primitives and proving propositions by deduction, is called *axiomatization* and has become the classic method of mathematics.

In geometry there are no causes, only inviolable relationships. Spinoza believed that the same was true of human beings. He approached what he called "the affects"—essentially all human emotions—the way Euclid approached triangles and squares, aiming to understand their interrelations by means of principles, logic, and deduction. His larger aim, culminating in part 5 of *Ethics,* entitled "Of the Power of the Intellect, or On Human Freedom," was to find a method to escape the violent sway of emotions on human beings caught in their grip.

*Ethics* is more a worldview than a tightly reasoned argument. Nevertheless I call what Spinoza created a theory. He makes no analogies; his primitives live in their own space; he doesn't attempt to explain how humans behave by comparing them to some other system. He employs observation, experience, introspection, and intuition to those human experiences familiar to everyone.

**The Primitive Sensations**

Spinoza's primitives are *pain, pleasure,* and *desire.* Every adult with a human body knows by direct experience what these sensations are, though Spinoza, following Euclid, attempts to define them. He begins with desire, which he terms man's essence, all of "man's strivings, impulses, appetites and volitions . . . which are not infrequently so opposed to one another that the man is pulled in different directions and knows not where to turn." "Desire," he summarizes, "is appetite conscious of itself." The consequences of desire are pleasure and pain:

> Pleasure is the transition of a man from a less to a greater perfection.
> Pain is the transition of a man from a greater to a less perfection.

Note that perfection is *not* pleasure and imperfection is *not* pain. Pleasure and pain are associated with transitions *between* states, not the states themselves. These seventeenth-century insights are in agreement with recent studies in psychology that find that people who obtain something they always wanted quickly become accustomed to their new possession or status and are soon no longer satisfied with it.

For now, assume that we know what we mean by *pleasure, pain,* and *desire.* They lie beneath all the other emotions and can conveniently be thought of as closer to organic conditions than psychic ones. That pain is fundamental is borne out by the fact that doctors test comatose patients for signs of life by looking for a response to pain. It is, of course, much more difficult to cause pleasure than to cause pain. Nevertheless, for Spinoza, pain and pleasure bear equal weight and can exist independently of each other; the absence of one doesn't indicate the presence of the other.

Spinoza distinguishes precisely between local and global sensations. "Pleasure and pain," he writes, "are ascribed to a man when one part of him is affected more than the rest, whereas *cheerfulness* and *melancholy* are ascribed to him when all are equally affected." *Suffering*, therefore, is localized pain, while *melancholy* is globalized pain.

His definitions of *good* and *bad* are pragmatic: "By good I here mean every kind of pleasure . . . especially that which satisfies our longings, whatsoever they may be. By evil, I mean every kind of pain, especially that which frustrates our longings." Good brings pleasure, and bad brings pain. Moral good is identical to sensual good; it's good to feel good and it's bad to feel bad.

**The Derivative Emotions**

Just as the value of a stock option depends on the underlying stock price, so the more complex human emotions depend, via one or more degrees of separation, on the three underlying primitive sensations. Spinoza's entire theory resembles the structure of contingent claims in modern finance, whose Efficient Market Model I describe in chapter 5. The following are paraphrases of some of Spinoza's definitions.

*Love* is pleasure associated with an external object. It is an emotion one step removed from simple pleasure.
*Hate* is pain associated with an external object.
*Hope* is the expectation of future pleasure when the outcome is uncertain and doubtful.
*Joy* is the pleasure we experience when that doubtful expectation is fulfilled.
*Disappointment* is the pain opposed to joy.

*Pity* is pain arising from another's hurt. Expanding on this, he writes,
"Pity is pain accompanied by the idea of evil, which has befallen
someone else whom we conceive to be like ourselves."

*Pity* involves two evils: the pain we experience as well as someone
else's pain. Correspondingly there should be an emotion involving
two goods: our own pleasure and someone else's. "What term we can
use for pleasure arising from another's gain, I know not," Spinoza
writes, adding that there are many derivative emotions that don't yet
have names.[5]

Like convertible bonds, modern-day corporate securities that
have both debt and equity characteristics, there are emotions that
depend on two underlying primitives. Thus *envy* is pain at another's
pleasure. Conversely, though the concept appeared long after Spinoza's death, *Schadenfreude* is pleasure at another's pain.

*Cruelty* links all three primitives: Spinoza defines it as the desire to
inflict pain on someone we love or pity.[6] Financially speaking, cruelty is analogous to a convertible bond whose debt and equity depend
on three economic underliers: the stock price, the level of interest
rates, and the credit worthiness of the company's debt.[7]

I call Spinoza's analysis a theory because it is an attempt to
describe the nature of emotions rather than compare emotions to
something else. You can get a feel for the scope and thoughtfulness
of his categorization by looking at a few more of his many subtle
decipherings:

We will call the love towards him who confers a benefit on another,
Approval; and the hatred towards him who injures another, we will
call Indignation.

Honour is pleasure accompanied by the idea of some action of our
own, which we believe to be praised by others.

Regret is the desire or appetite to possess something, kept alive by the remembrance of the said thing, and at the same time constrained by the remembrance of other things which exclude the existence of it.

I shall call him intrepid who disdains an evil I usually fear.

Repentance is sadness accompanied by the idea of oneself as cause, and self-esteem is joy accompanied by the idea of oneself as cause. Because men believe themselves free, these affects are very violent.

## Three Meta-affects

To span the complexity of human emotions, Spinoza adds to his theory three additional primitives that are closer to meta-affects than affects themselves. The first is *vacillation,* a state of oscillation between two emotions. Thus *jealousy,* he explains, is the vacillation between hate and envy toward an object of love in the presence of a rival. Jealousy depends on envy, and envy, as we have seen, depends on pleasure and pain. If we follow the links far enough, every affect terminates at pain, pleasure, or desire.

The second meta-affect is *wonder.* Wonder is what we experience when confronted by something that fills the mind to the exclusion of all else, something unrelated to anything else we know. Wonder is what Moses experienced at the burning bush, in the presence of Yahweh, who is what He is. Spinoza expands on wonder:

> But if it be excited by an object of fear, it is called Consternation, because wonder at an evil keeps a man so engrossed in the simple contemplation thereof, that he has no power to think of anything else whereby he might avoid the evil. . . .
>
> Otherwise, if a man's anger, envy, &c., be what we wonder at, the emotion is called Horror. . . .

> If it be the prudence, industry . . . of a man we love, that we wonder at, our love will on this account be the greater, and when joined to wonder or veneration is called Devotion.

I will illustrate wonder in chapter 4, when I recount the miraculous development of quantum electrodynamics, a theory that exemplifies better than anything else the triumph of the mental over the physical.

Spinoza's final meta-affect is *contempt,* the feeling we have when we contemplate something that most forcibly reminds us of the qualities it lacks, their absence a palpable presence. He links contempt to *derision* and *scorn*:

> As devotion springs from wonder at a thing which we love, so does Derision spring from contempt of a thing which we hate or fear, and Scorn from contempt of folly, as veneration from wonder at prudence. Lastly, we can conceive the emotions of love, hope, honour, &c., in association with contempt, and can thence deduce other emotions, which are not distinguished one from another by any recognized name.

Wonder, vacillation, and contempt lie not so much beneath all affects as to the side of them.

I have created a map of Spinoza's verbal definitions in Figure 3.1. At the center of the diagram is the primitive *pain.* The solid black arrows emanating from *pain* lead to its derivative affects. To the lower right is *pleasure,* whose dashed arrows lead to pleasure-related emotions. Thus *pride* is the pleasure derived from thinking too highly of oneself. In the upper right is *desire,* the driver of all action, whose heavy black arrows of cupidity point to its dependents. Thus *emulation* is "the desire of something engendered in us by our conception that others have the same desire," a type of inauthenticity in that the cause of our desire is someone else's desire. Mass-market advertising depends on it.

# Pleasure, Pain, Desir

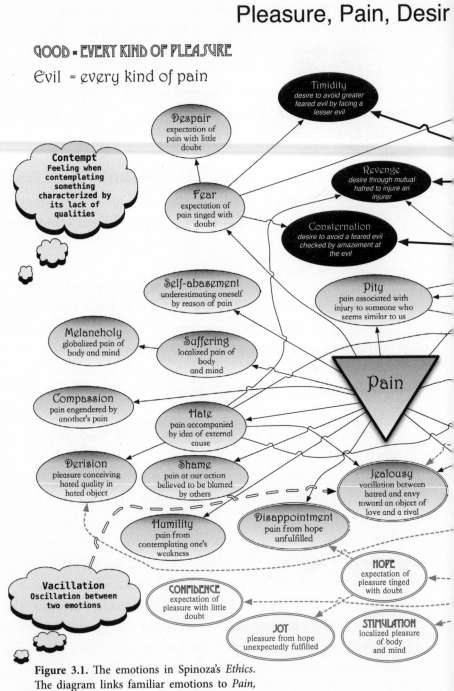

**GOOD = EVERY KIND OF PLEASURE**

**Evil** = every kind of pain

**Timidity**
*desire to avoid greater feared evil by facing a lesser evil*

**Despair**
expectation of pain with little doubt

**Contempt**
Feeling when contemplating something characterized by its lack of qualities

**Revenge**
*desire through mutual hatred to injure an injurer*

**Fear**
expectation of pain tinged with doubt

**Consternation**
*desire to avoid a feared evil checked by amazement at the evil*

**Self-abasement**
underestimating oneself by reason of pain

**Pity**
pain associated with injury to someone who seems similar to us

**Melancholy**
globalized pain of body and mind

**Suffering**
localized pain of body and mind

**Pain**

**Compassion**
pain engendered by another's pain

**Hate**
pain accompanied by idea of external cause

**Derision**
pleasure conceiving hated quality in hated object

**Shame**
pain at our action believed to be blamed by others

**Jealousy**
vacillation between hatred and envy toward an object of love and a rival

**Humility**
pain from contemplating one's weakness

**Disappointment**
pain from hope unfulfilled

**HOPE**
expectation of pleasure tinged with doubt

**Vacillation**
Oscillation between two emotions

**CONFIDENCE**
expectation of pleasure with little doubt

**JOY**
pleasure from hope unexpectedly fulfilled

**STIMULATION**
localized pleasure of body and mind

**Figure 3.1.** The emotions in Spinoza's *Ethics*. The diagram links familiar emotions to *Pain, Pleasure,* and *Desire*.

# Map of the Emotions

**Wonder**
Feeling when contemplating something unconnected to everything else

**Daring**
desire to face a danger which equals fear

**Cowardice**
desire checked by fear of a danger which other equals dare to face

**Courtesy**
desire to act so as to please people and refrain from displeasing them

**Regret**
desire to possess something tinged by awareness of other things which prevented the possession

**Cruelty**
desire to injure someone we pity or love

**Desire**

**Emulation**
desire engendered by another's desire

**Injurer**
one who brings pain

**Anger**
desire to injure hated one

**Indignation**
hatred towards an injurer

**Benevolence**
desire to benefit one whom we pity

**Gratitude**
love for and desire to benefit a loving benefactor

**Envy**
pain at another's pleasure

**LOVE**
pleasure accompanied by idea of external cause

**APPROVAL**
love toward a benefactor

**Schadenfreude**
pleasure at another's pain

**DEVOTION**
love toward one we admire

**BENEFACTOR**
one who brings pleasure

**SYMPATHY**
pleasure/pain from another's pleasure/pain

**PLEASURE**

**SELF-APPROVAL**
pleasure from contemplating one's strength

**MERRIMENT**
globalized pleasure of body and mind

**HONOR**
pleasure at our action believed to be praised by others

**PRIDE**
pleasure from thinking too highly of oneself

**DISDAIN**
pleasure from thinking too little of someone

In the map you can examine the link between the hybrid emotions and the primitives. *Courtesy* is the desire to please people and to refrain from displeasing them, linked to both *pleasure* and *desire*. *Revenge* is the *desire* through mutual hatred to bring *pain* on someone who has brought us *pain*. *Regret*, the saddest of emotions, is the desire to possess something, tinged with the recollection of a past that made it impossible. Who hasn't known it?

## FIAT MONEY

I have used Spinoza's framework to look at the idea of money, a topic fraught with a variety of emotions.

Once upon a time money was gold coinage; later it was paper representing a claim on gold deposits; nowadays it's fiat money, a medium of exchange and value that is purely conventional, freely created, and anchored by nothing except authority. "Money," wrote Schopenhauer, looking a little deeper, "is human happiness in the abstract; he then who is no longer capable of enjoying human happiness in the concrete devotes himself utterly to money."

But genuine money is more than crystallized pleasure. It doesn't come easy, except to those who print it. "By the sweat of your brow will you eat bread," said God to Adam and Eve after the Fall. Genuine money is crystallized work, and work is pain in the service of the desire to survive. Genuine money is also security, the expectation of future pleasure and freedom from pain. It combines in one object all three of Spinoza's primitives, as illustrated in Figure 3.2.

Fiat money, unlike genuine money, incorporates only pleasure and desire. Without work, it lacks the connection to pain that gives it value and respect. By virtue of this lack, it induces contempt.[8]

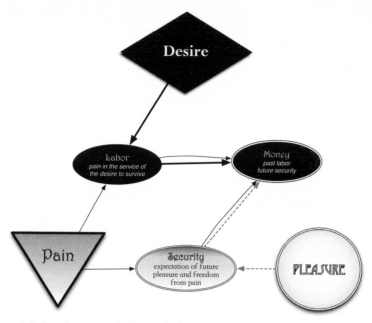

**Figure 3.2.** Genuine money in Spinoza's theory

## LOVE AND DESPERATION

It seems to me that love can be more complex than Spinoza's simple pleasure associated with an external object. Sometimes love can be the desperate pleasure of giving in to desire and abandoning control. Madame Bovary, Humbert Humbert, Charles Swann, Anna Karenina—for all of them love was mingled pain and pleasure. This kind of love is as much the longing for the cessation of the pain of longing as it is the longing for the commencement of pleasure. It vacillates between desire, hope, and despair. It is fibrillation, and the love of it is a death wish.

But people seem to need the desperate side of love. The citizens of Aldous Huxley's *Brave New World* had easy access to unemotional physiological lovemaking. As a result every few years they had to undergo a VPS (Violent Passion Surrogate). The synthetic passions of a periodic biochemical love affair were necessary to keep their bodies functioning smoothly.

## HOW TO LIVE IN THE REALM OF THE PASSIONS

Spinoza spent most of his life working on his *Ethics*. His theorizing was aimed at practice; he intended it to provide a cure for the passions and a practical blueprint for living. The word *passion* originates from the Latin word for suffering, as in the Passion of Christ. Passion sweeps across us from the outside, destroying our capacity to reason. Spinoza's cure for the bondage of the passions is to understand ourselves both from within and from without.[9] The result will be action based on understanding rather than passion driven by emotions. Understanding leads to autonomous action, the capacity to be an underlier rather than a derivative.

Parts of *Ethics* are so cryptic as to seem impenetrable. But, as Gilles Deleuze writes:

> There is a double reading of Spinoza; on the one hand, a systematic reading in pursuit of the general idea and the unity of the parts, but on the other hand and at the same time, the affective reading, without an idea of the whole, where one is carried along or set down, put in motion or at rest, shaken or calmed according to the velocity of this or that part.[10]

What follows is *my* affective reading, riven with complications, discrepancies, and contradictions, which I hope will seem reconciled by chapter's end.

## THE FOUR QUESTIONS

### 1. Why Do We Treat Ourselves Differently from Others?

I tend to treat other people as though they are responsible for their deeds and misdeeds. When they do something that hurts me, I regard them as free agents who could have done otherwise. But what about the harmful things I do? In my case I often feel as though I'm in the grip of tidal pulls that make me do things I wish I hadn't done. I tend to excuse myself with the understanding that I can't help it.

Observing others from outside, I hold them responsible. Observing myself from inside, I can always think of good excuses for my behavior. In their case it's Will; in my case I call it circumstance. Why one explanation for their behavior, another for mine?

### 2. How Can We Control Ourselves?

We are filled with desires whose origins we don't understand. When we act on them, we imagine we are behaving freely. But as Spinoza wrote:

> Men are conscious of their own desire, but ignorant of the causes whereby that desire has been determined.
>
> But experience teaches all too plainly that men have nothing less in their power than their tongue, and can do nothing less than moderate their appetites. . . . So the madman, the chatterbox, the child, and a great many people of this kind believe they speak from a free decision of the mind, when really they cannot contain their impulse to speak.

Our apparent freedom is only the freedom to do what we want. Our volition drives us. What drives our volition? "You can do what you want, but you cannot want what you want," wrote Schopenhauer on the same topic.

Madame Bovary, Anna Karenina, Humbert Humbert—couldn't they have stopped themselves?

### 3. What Is the World Made Of?

Since Descartes or even before, we have regarded some parts of the world as matter and other parts as mind. Why are there two apparently distinct substances? What, if any, is the connection between them?

### 4. Why Does the Material World Have Laws While the Human World Has Explanations?

Matter satisfies Newton's laws, Maxwell's equations, the Dirac theory of the electron. Matter has no freedom of action. You can't ask the falling ball "Why?" The human world, on the other hand, has explanations for its behavior:

> Why did you hang up on me?
> *Because you insulted me. Why can't you at least be civil?*
> Because you enjoy provoking me. Why do you take everything to extremes?
> *Because . . .*

Such explanations can be endlessly recursive, searching hopelessly for a first cause.

## SPINOZA'S ANSWERS

Spinoza divided knowledge of the world into three categories: adequate knowledge, inadequate knowledge, and intuition.

### ADEQUATE KNOWLEDGE

When my son was a little less than two years old, I used to play a game he liked, bouncing him up and down on my knee while chanting a nursery rhyme:

> *Half a pound of tuppeny rice*
> *Half a pound of treacle*
> *Mix them up and make them nice*
> *Pop goes the weasel!*

On "Pop goes the weasel!" I would sharply drop my knee all the way down and let him bump to the floor. He always chortled. He liked the game so much that one day he asked me to repeat it over and over again, laughing at each bump except the final one, when he turned to me in surprise and asked, "Why it's not funny anymore?" From outside himself, he understood something within himself. He had discovered that something repeated over and over again becomes progressively less funny until it's not funny at all. This is an example of *adequate knowledge*.

Adequate knowledge is an apprehension that is self-contained, that leans on nothing else. The Dirac equation, the theory of evolution, Freud: these are adequate explanations. There is no need to ask "Why?" when something is adequately explained. The explanation is sufficient; the theory is the fact. Dirac discovered that electrons satisfy the Dirac equation. My son discovered that funniness fades with repetition. That's how God's world works.

Adequate knowledge is global: it transcends a single individual or a singular occurrence. It is always-true knowledge rather than ad hoc knowledge. Adequate knowledge is a comprehension of relationships rather than causes. In Spinoza's words:

> I call that cause adequate whose effect can be clearly and distinctly perceived through it. But I call it partial, or inadequate, if its effect cannot be understood through it alone.

Theories are adequate knowledge. Models are inadequate.

**INADEQUATE KNOWLEDGE**

When we truly understand an occurrence, we have *adequate* knowledge of it; when we don't, when we are unable to explain an occurrence in generality, we have *inadequate* knowledge. For example:

> The financial crisis of 2007–2008 was caused by the global savings glut.
> *What caused that?*
> The Asian currency crisis of 1997–1998: Asian countries came out of that wanting to run net surpluses rather than net deficits.
> *But what caused that?*
>
> . . .

Each explanation, reasonable though it sounds, provokes a request for another. Each explanation is inadequate and local because it displaces the ultimate cause one degree further from the final effect.

## Extending the Scope of Adequate Knowledge

As time passes we understand more things adequately. To take an example given by Stuart Hampshire in his book on Spinoza, humans may once have thought that lightning indicated that the gods must be angry; now we know that lightning is caused by a difference in electric potential, as described by Maxwell's equations. In similar fashion, while people used to cite "evil" as the mysterious source of criminal behavior, we now often blame it on "parental neglect."[11] In this way we progress from inadequate understanding in terms of the arbitrary volition of the gods to understanding via more general laws.

Similarly, claims Spinoza, human volition is an inadequate explanation for human actions, and it must be replaced by a deeper understanding via laws of human behavior: "To conceive a thing as free can be nothing else than to conceive it simply, while we are in ignorance of the cause whereby it has been determined to action." Saying we have acted freely is tantamount to saying we don't understand why we acted. When we follow our passions "freely" we are in fact behaving deterministically, according to the well-known laws of physics and the poorly known laws of mind. We understand the laws of matter well, and know that it is futile to complain about the behavior of electrons. Man's behavior is also subject to laws. The better we understand them, the more tolerant we will be. The Dirac equation is a more profound explanation of reality than "I did this because you did that."

Hegel wrote that "the history of the world is none other than the progress of the consciousness of freedom." Spinoza's claim is that the history of the world is none other than the progress of the consciousness of our lack of freedom.

If there are laws of the universe governing human behavior, how can we learn them?

## II. MODELS BEHAVING

### From the Particular to the General, from Reason to Intuition

According to Spinoza, there are a variety of ways to attain understanding of the world:

- Via particulars
- Via generalities
- Via intuition

Particulars are diverse and confusing, but they provide the basis for understanding everything else: "The more we understand particular things, the more do we understand God,"[12] that is, Nature, in its entirety. Applying reason to particulars, we obtain adequate ideas about the regularities common to all things. A step beyond that, according to Spinoza, is the deepest kind of knowledge:

The highest endeavor of the mind, and the highest virtue is to understand things by the intuitive kind of knowledge.

#### INTUITION

It takes intuition to discover theories. Intuition may sound casual, but it emerges only from intimate knowledge acquired after careful observation and painstaking effort. Before you can move one level higher in the pyramid of understanding, before you can attain intuition in some domain, you have to struggle with the particulars of that domain until knowledge of its details is second nature to you.

A cyclist develops physical intuition about the correct angle to tilt body and bicycle to a curved track so as to maximize stability; the builder of a velodrome can calculate the correct banking angle to ensure the cyclist remains in equilibrium; together biker and builder combine visceral and theoretical knowledge. Intuition is learning to ride a bicycle without thinking. You have to incorporate the laws of the world into your body.

Feynman's insight into the parallel evolution of quantum mechanical paths, Dirac's grasp of the essence of electrons, Newton's understanding of mass and its motion—all are instances of the external world joining with the internal. Intuition is a merging of the understander with the understood. In the words of the Upanishads, *Tat tvam asi*, Thou art that.

## Perfection via Intuition

"Pleasure is the transition of a man from a less to a greater perfection," wrote Spinoza. No one achieved greater advances in our levels of perfection than Isaac Newton, born in 1642, only ten years later than Spinoza. John Maynard Keynes wrote a speech about Newton for the tercentenary of his birth, celebrated belatedly by the Royal Society in 1946, after World War II. By then Keynes had died, and his brother Geoffrey delivered the speech. It was based on Keynes's reading of a box of Newton's notes, many of them cryptic and mystical, concerning his attempts to understand not just the physical but the entire world.

> Newton came to be thought of as the first and greatest of the modern age of scientists, a rationalist, one who taught us to think on the lines of cold and untinctured reason. I do not see him in this light. Newton was not the first of the age of reason. He was the last of the magicians, the last of the Babylonians and Sumerians, the last great mind which looked out on the visible and intellectual world with the same eyes as those who began to build our intellectual inheritance rather less than 10,000 years ago. . . .
>
> I believe that the clue to his mind is to be found in his unusual pow-

ers of continuous concentrated introspection. . . . His peculiar gift was the power of holding continuously in his mind a purely mental problem until he had seen straight through it. I fancy his pre-eminence is due to his muscles of intuition being the strongest and most enduring with which a man has ever been gifted. Anyone who has ever attempted pure scientific or philosophical thought knows how one can hold a problem momentarily in one's mind and apply all one's powers of concentration to piercing through it, and how it will dissolve and escape and you find that what you are surveying is a blank. I believe that Newton could hold a problem in his mind for hours and days and weeks until it surrendered to him its secret. Then being a supreme mathematical technician he could dress it up, how you will, for purposes of exposition, but it was his intuition which was pre-eminently extraordinary—"so happy in his conjectures," said De Morgan, "as to seem to know more than he could possibly have any means of proving."

There is the story of how he informed Halley of one of his most fundamental discoveries of planetary motion. "Yes," replied Halley, "but how do you know that? Have you proved it?" Newton was taken aback— "Why, I've known it for years," he replied. "If you'll give me a few days, I'll certainly find you a proof of it"—as in due course he did. . . .

Certainly there can be no doubt that the peculiar geometrical form in which the exposition of the *Principia* is dressed up bears no resemblance at all to the mental processes by which Newton actually arrived at his conclusions.

His experiments were always, I suspect, a means, not of discovery, but always of verifying what he knew already.

Why do I call him a magician? Because he looked on the whole universe and all that is in it as a riddle, as a secret which could be read by applying pure thought to certain evidence, certain mystic clues which God had laid about the world to allow a sort of philosopher's treasure hunt to the esoteric brotherhood. He believed that these clues were to be found partly in the evidence of the heavens and in the constitution of elements (and that is what gives the false suggestion of his being an experimental natural philosopher), but also partly in certain papers and traditions handed down by the brethren in an unbroken chain back to the original cryptic revelation in Babylonia. He regarded the universe as a cryptogram set by the Almighty—just as he himself wrapt the discovery of the calculus in a cryptogram when he communicated with Leibniz. By pure thought, by concentration of mind, the riddle, he believed, would be revealed to the initiate.

He did read the riddle of the heavens. And he believed that by the same powers of his introspective imagination he would read the riddle of the Godhead, the riddle of past and future events divinely foreordained, the riddle of the elements and their constitution from an original undifferentiated first matter, the riddle of health and of immortality. All would be revealed to him if only he could persevere to the end, uninterrupted, by himself, no one coming into the room, reading, copying, testing—all by himself, no interruption for God's sake, no disclosure, no discordant breakings in or criticism, with fear and shrinking as he assailed these half-ordained, half-forbidden things, creeping back into the bosom of the Godhead as into his mother's womb. "Voyaging through strange seas of thought alone," not as Charles Lamb "a fellow who believed nothing unless it was as clear as the three sides of a triangle."

Keynes saw Newton's explanations as merely a means of verifying what he had already discerned more directly. This was also James Clerk Maxwell's impression of André-Marie Ampère, who in 1820 discovered the connection between electricity and magnetism. Referring to him as the "Newton of electricity," Maxwell wrote:

We can scarcely believe that Ampère really discovered the law of action by means of the experiments which he describes. We are led to suspect, what, indeed, he tells us himself, that he discovered the law by some process which he has not shown us, and that when he had afterwards built up a perfect demonstration, he removed all traces of the scaffolding by which he had built it.

For Spinoza, God's understanding of the world is like that of the cyclist banking on a curve, perfectly stable without his having to think, in unison with the road. God's is the ultimate intuition, so encompassing that there is no boundary between the creator and the created.

**PERFECTION**

Pleasure is the transition to a greater perfection, but what is perfection itself? Early in *Ethics,* Spinoza defines it: "Reality and perfection I use as synonymous terms."

How can reality be perfection? Our world is not a Panglossian best of all possible worlds. Nevertheless, I observe that our language reveals a sense in which the real and the perfect are synonymous:

*Perfect means completed:* The perfect tense in grammar refers to actions that have been completed.

*Completed means successfully realized:* Only that which has been completed is actual and realized. (A quarterback "completes a pass.")

*Therefore perfect means real:* Only that which has been completed as its author intended, whose realization matches the mental model that preceded it, is perfect.

**A True Theory Is Perfection**

An idea, as we will shortly see, is always inseparably associated with matter; each is a side of one larger thing. Thus truth, according to Spinoza, is a harmony between an object and its idea.

*An idea that corresponds with its object* is a very good definition of a correct theory.

## Levels of Perfection

Spinoza understood that our fundamental desire is to remain integral, to survive: "No virtue can be conceived prior to the virtue of striving to preserve oneself." Pleasure and goodness are conducive to survival, while pain and evil promote disintegration, a change of state. As Spinoza writes:

> The emotion of pain is an . . . activity of transition from a greater to a less perfection. In other words, it is an activity whereby a man's power of action is lessened or constrained.

Spinoza is not naïve about pleasure; he knows that it is not an unmitigated good, and that it is best when it is balanced. He distinguishes pleasure from stimulation (Latin *titillatio*), a pleasure that is focused more on some parts of the body than others. Localized pleasure, he writes, can exert an obsessive power "that can overcome other actions of the body, and may remain obstinately fixed therein, thus rendering it incapable of being affected in a variety of other ways: therefore it may be bad." Who could disagree?

Because stimulation accompanied by the idea of an external cause is, by his definition, a kind of love, love may be excessive too. Generalized pleasure, though, is always good:

> Mirth is pleasure, which . . . consists in all parts of the body being affected equally: that is, the body's power of activity is increased or aided in such a manner, that the several parts maintain their former proportion, therefore Mirth is always good, and cannot be excessive.

SUBSTANCE, MIND, AND MATTER

Everything we are aware of manifests itself as either matter or mind. In Spinoza's view, these manifestations are the features of only one Substance. Substance has many attributes, but we humans can perceive only two of them: a mind-quality (Thought) and a matter-quality (Extension). The word *substance* originates in the Latin *substare,* "to stand beneath." In the language of finance, substance is the ultimate underlier, and everything else is its derivative.

Mind and matter are simultaneous attributes of Substance. Mind is not an epiphenomenon of matter, nor is matter an epiphenomenon of mind. Neurophysiology doesn't explain psychology, and psychology doesn't replace neurophysiology. Both are different views of the same underlier.

Things made out of Substance obey the deterministic laws of the universe, but it's not easy to deduce those general laws when you observe only particulars. Lest you think it naïve to assume that there are laws behind everything, recall how many centuries of observing the lights in the night sky it took to discover that Newton's three laws of motion and his law of gravity could explain the motions of the planets, the stars, and objects on earth. Tycho Brahe had to map the planetary motions and Johannes Kepler had to intuit that they described mathematical ellipses, each planet sweeping out equal areas in equal times; only then could Newton step into the picture with dynamics.

With time, what more may we still discover?

**Mysterious Materialism**

Spinoza is a materialist, but not a naïve materialist. Since Newton we think of matter as dull, inanimate stuff that must obey laws. But Spinoza points out that the matter we inhabit is full of mysterious possibilities:

> However, no one has hitherto laid down the limits to the powers of the body, that is, no one has as yet been taught by experience what the body can accomplish solely by the laws of nature. . . . Nor need I call attention to the fact that many actions are observed in the lower animals, which far transcend human sagacity, and that somnambulists do many things in their sleep, which they would not venture to do when awake: these instances are enough to show, that the body can by the sole laws of its nature do many things which the mind wonders at.

The brain, after all, is part of the body too. One of the lessons of twentieth-century physics—of relativity, quantum mechanics, and cosmology—is that the more we learn about matter, the more enigmatic it seems.

There is a "mind" way of looking at things and there is also a "matter" way. In his book *I Am a Strange Loop*, Douglas Hofstadter imagines a digital computer built out of chains of dominoes constructed to divide the prime number 641 by all the numbers less than it. You begin the program by knocking over the first domino. The logic of the chains is such that if no number can divide 641 without a remainder, then the final domino in the chain will fall. The domino computer begins its computation, and the final domino falls a few seconds later. *Why did the domino fall?* Answer 1: Because the domino preceding it in the chain pushed it over. *And why did that domino fall?* Because the domino preceding it fell and pushed it over. Et cetera. But there is also Answer 2: Because 641 is a prime number. Both of these answers are simultaneously true, the first in the realm of matter, the second in the realm of mind. The dominoes don't know about primes, and primes don't know about dominoes.

Each explanation is separately valid. But the logical chains that drive each explanation—falling dominoes in the matter realm, division by primes in the mind realm—cannot be independent of each other. It makes no sense, says Spinoza, to have two different causes for the same single sequence of events. Instead, according to Spinoza, there is one unique sequence of events, and there is one unique causal chain that accounts for both explanations.

Stuart Hampshire sharpens this argument and thereby makes it even harder to accept. Suppose you become embarrassed and turn red. You might say, "I blushed because I became embarrassed." A strict Spinozist would not claim that embarrassment was the cause of blushing, because embarrassment is the *mental description* of the *physical* blush, a crisscrossing of causal chains. We should not jump from one style of explanation to another. We must explain physical things by physics and psychological things by psychology. It is of course very difficult to give up the notion of psychic causes for physical states. But, as Spinoza says, no one knows by what means the mind moves the body.

*Ethics* argues that every manifestation of Substance must appear as both Thought and Extension. (In Hebrew, the word for *word* is the same as the word for *thing*.) Mind and the emotions, like matter, are not extraordinary; they lie within, not outside, Nature and its laws. From this point of view, the body and the mind are reciprocal: "The body is the object of the mind. The object of the idea constituting the human mind is the body."

It is tempting to assume that there is an idea corresponding to everything, and hence, recursively, an idea of an idea too. To which Spinoza preemptively answers, "This idea of the mind is united to the mind in the same way as the mind is united to the body." That is, the idea of the mind is the mind itself. The circle closes.

## Emotions Are the Link

Why does Spinoza place so much importance on the emotions? Because mind and body proceed in parallel disconnected paths, and emotions are the only perceptible link between them. The passions are the wormhole between the two sides of our little piece of dream-stuff:

> The human mind has no knowledge of the body, and does not know it to exist, save through the ideas of the modifications whereby the body is affected.
>
> The mind does not know itself, except in so far as it perceives the ideas of the modifications of the body.

By the "modifications of the body" Spinoza means the affects or emotions. The only way the mind can know the body is through the emotions, which tie the two spheres together. The *idea* side of our physical responses to external interactions are the emotions, hence their critical importance.

Though the dominoes don't know about primes, and the primes don't know about dominoes, the body and the mind know each other through the affects.

## Our Understanding of Our Body Is Unclear

Like most people, I function pretty well without understanding anything about my physiology. It may even be a distraction to know too much about one's internal structure. From Spinoza's viewpoint, this ignorance is to be expected: it is only through the passions that we can know our body, and since the passions are caused by interactions with other bodies, we cannot form a clear idea of our own body in isolation.

### THE CURE: UNDERSTANDING THE ADEQUATE CAUSES

Our passions, properly understood, are *re*actions to something outside us. If we had a clear idea of the laws that constrain us, we would adequately understand the causes of our passions. They are passions only as long as we don't understand them. Once we do, proceeding *with* understanding becomes an *action,* and we become an independent underlier rather than a dependent derivative.

We can, Spinoza claims, convert our passions into actions by understanding their true causes: "If we can be the adequate causes of any of these affections, I understand by the affect an action; otherwise a passion." Spinoza's cure is in keeping with contemporary notions: we need to lay bare the subconscious drivers of our feelings. When you understand yourself from inside and out, you know yourself. Freedom is the unification of understanding and volition, of reason and desire. Will and Understanding are one and the same. Understanding is merely Will perceived from the inside.

With this understanding, you become close to Spinoza's God, who does not *think* about what to do. He operates with intuition. He does not consider the possibilities and then do the right thing. There is no *need* for Him to act. He is complete and perfect. He does what He does and He is what He is. He's the Understander and the Understood. He's not a metaphor.[13]

# CHAPTER 4

# THE SUBLIME

*The birds of the air • The best theory in the world • No logical path to it • Electricity and magnetism • Their qualities • Their quantitative laws • Ampère's sympathetic understanding of the phenomena • Faraday's imaginary lines of force • Maxwell's factual field • Dodging the beasts of the field • Quantum dreams*

There is a delicate empiricism which makes itself utterly identical with the object, thereby becoming true theory. But this enhancement of our mental powers belongs to a highly evolved age.

—Goethe, *Maxims and Reflections*

One of the points I have laboured in this book is the unitary source of mystical and scientific modes of experience.

—Arthur Koestler, *The Sleepwalkers*

The ultimate goal would be: to grasp that everything in the realm of fact is already theory.

—Goethe, *Maxims and Reflections*

## THE BIRDS OF THE AIR

In his essay "Frogs and Birds," the introduction to a collection of papers by the Russian mathematician Y. Manin, Freeman Dyson wrote:

> Some mathematicians are birds, others are frogs. Birds fly high in the air and survey broad vistas of mathematics out to the far horizon. They delight in concepts that unify our thinking. . . . Frogs live in the mud below and see only the flowers that grow nearby. They delight in the details of particular objects, and they solve problems one at a time. I happen to be a frog, but many of my best friends are birds.

Dyson's metaphorical classification of mathematicians applies to physicists too. I was mostly a clumsy sort of frog physicist, elaborating or testing other people's ideas, but I was a bird manqué. I studied the work of birds, among them Newton, Ampère, Maxwell, Einstein, Schrödinger, Feynman, Gell-Mann, and Weinberg—men who had intuited and discovered wonderful, astonishing things about the world. Perhaps I could do it too. Deep inside, every physicist dreams of glory and believes it is attainable, or once did so; otherwise he or she wouldn't be in the field.

As long as I was immersed in doing physics, the discovery of its laws seemed a natural and obvious process. Now when I look at the field and realize that I can no longer recall the origin or proof of some facts I once knew as second nature, I am awed that anyone was ever able to penetrate through the phenomena to the laws. In a speech on the principles of research, given in 1918 in honor of Max Planck, the discoverer of the quantum, Einstein captured the mystery of the birds' accomplishments: "There is no logical path to these laws; only intuition, resting on sympathetic understanding of experience, can reach them."

Of all theories, the best in the world is quantum electrodynamics, appropriately called QED. QED is the quantum theory of the electron and its interaction with light, and it determines just about everything relevant to the physics and chemistry of the atoms and molecules that compose us and the world around us. We trust it because it predicts the values of measured properties of the atom so precisely as to strain belief. Any successful theory in physics is amazing; QED is a miracle.

As I'll show in the next chapter, the creation of QED runs in counterpoint to the development of the Efficient Market Model in finance. The development of QED is a tight intertwining of data, facts, experiments, failures, and successes; the development of the Efficient Market Model is coupled to the world much more loosely, driven as much by ideology as by facts. QED predicts atomic properties to an accuracy of more than 10 significant figures; the best models in finance are not accurate to even one. More to the point, no one in finance knows how to specify exactly what "accurate" means, because so many of its variables are related to human sentiment.

In this chapter I want to provide a glimpse of the crooked paths that culminate in theories, to give a sense of the role of intuition in the discovery of a theory and in the sweep of that discovery. I want to recount in stylized fashion some of the prodigious feats that birds and frogs have achieved as they created the best theory in the world. Here, then, is an attenuated trip through the history of classical and quantum electromagnetic theory, a triumph of mind over matter.

## THE PHENOMENA:
## ELECTRICITY AND MAGNETISM

First the facts. The ancient Greeks knew that rubbing amber, the fossilized tree resin called *elektron* in Greek, empowered it to pick up tiny scraps of straw or feather. We now call this *static* electricity, the buildup of a stationary electric charge on the surface of objects, a terminology that prefigures the later discovery that what we call an electric current is the *dynamic* flow of charge. The Greeks were also familiar with lodestones, naturally occurring magnets that they discovered in Magnesia in the province of Thessaly.

Then the theories. Lodestones pointed north, but why? Some thought the attractor was the polestar in the heavens; others imagined a small magnetic island located at the north pole itself. Then, in 1600, William Gilbert, an English physician, proposed that the Earth itself is a giant magnet with an iron core. Proof was his experiment on the angles at which freely suspended needles "dip" in the vicinity of a spherical lodestone, which he noticed was similar to the way compass needles incline to the surface of the Earth. His theory is now a fact.

## QUALITIES: POSITIVE AND NEGATIVE

Yet electromagnetic theory was slow to take flight. More than 50 years after Newton began applying the calculus to the motion of the planets, at about the same time (1738) that Daniel Bernoulli published his seminal book on hydrodynamics, the theory of electricity and magnetism was just beginning to incorporate facts.

That there are two kinds of charge was a discovery made by the French chemist Charles Du Fay in 1734. He found that he could create two different types of electricity by friction: *vitreous* by rubbing glass-like materials, and *resinous* by rubbing resin-like materials. Furthermore the small pieces of charged glass that repelled each other attracted pieces of charged amber. Some years later Benjamin Franklin replaced Du Fay's empirically inspired adjectives by the moral descriptors *positive* and *negative*.

Note that electrical charge didn't *have to* come with a positive or negative sign. In gravitation everything is positively charged and masses never repel each other. The electron's antiparticle, even though it has the opposite electric charge, has the same gravitational mass. Masses carry no + or −.

## QUANTITIES: COULOMB'S LAW OF FORCE
## BETWEEN STATIC CHARGES

In 1788, inspired by Newton's law of gravitation, Charles Coulomb published what is now called Coulomb's law: the attractive or repulsive force between small point-like charges is proportional to the magnitude of their charge and diminishes in proportion to the inverse square of the distance between them. But take no discovery for granted: though it sounds simple, it took ingenious experimentation to accurately measure both the quantity of charge on objects and the forces between them.

No one thought much about the transmission of forces. People believed, even took for granted, that both the gravitational force and the electrostatic force were transmitted through space instantaneously by so-called action at a distance.

## VOLTA'S ITALIAN INSIGHT:
## CHEMISTRY IS BETTER THAN FRICTION

Facts accumulate. Until this moment the only way to produce electric charge was by means of friction, scraping charge off objects and storing it in small quantities in a reservoir, a Leyden jar for example. Electric currents fueled by such capacitors were consequently brief and transient. Then, in 1800, Alessandro Volta invented the first chemical source of electric charge, the *voltaic pile,* a battery consisting of alternating disks of copper and zinc separated by cardboard soaked in salt water.

It wasn't immediately obvious to anyone that what flowed out of the battery was electric charge, but its effects were soon identified with those caused by the brief flow of charge produced the old-fashioned, frictional way. Large voltaic piles produced powerful, steady, long-lived currents suitable for experimenting with. The discovery that running a current through water separated it into its component parts, hydrogen and oxygen, soon followed. Thus electrochemistry was born.

## OERSTED: ELECTRIC CURRENTS
## BEHAVE LIKE MAGNETS

More observations. Until now electricity and magnetism had seemed to be two disparate phenomena. Then, in 1820, Hans Christian Oersted discovered a crossing of boundaries: he observed that electric currents could deflect compass needles.

## AMPÈRE: A LAW FOR THE FORCE
## BETWEEN CURRENTS

If electric currents deflect compass needles, then currents behave like magnets, and must therefore interact with other magnet-like currents. Hearing of Oersted's discovery, André-Marie Ampère immediately began his own series of much more thorough and quantitative investigations.

First, he found that pairs of electric currents moving in the same direction attract, whereas pairs moving in opposite directions repel, as shown in Figure 4.1. Then, after using a voltaic pile to drive cur-

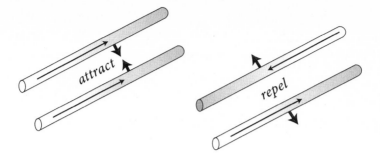

**Figure 4.1.** Ampère: Currents moving in the same direction attract. Currents moving in opposite directions repel.

rent through a helical loop of wire, he noticed that the force exerted on a compass needle by the loop was much like the force exerted by a magnet. He concluded that magnets themselves might be small solenoidal currents (see Figure 4.2)

This was qualitative, and relatively simple. Ampère's theoretical tour de force was his discovery of the mathematical formula for the magnetic force between two isolated *current elements,* tiny imaginary bits of current that are the magnetic analogues of point-like electric charges. I say "imaginary" because it is impossible to isolate small steady current elements. Currents from batteries can flow only in finite closed loops from one terminal to another.

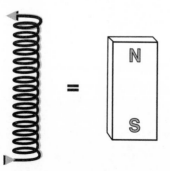

**Figure 4.2.** Ampère: Small current loops behave like little north-south bar magnets.

Ampère's law for the force between two imaginary bits of wire, as illustrated in Figure 4.3, was vastly more complex and subtle than the Coulomb force between two point charges, because charges have location but no orientation, whereas currents necessarily point in some direction. Ampère's force therefore varied not only with distance between the current elements but also with the orientation of each element, and is much more difficult to describe or visualize. Its discovery was a triumph of imagination and intuition.

Using his law Ampère was able to mathematically sum up the forces from each infinitesimal segment of current to determine the forces between entire circuits. His theory shifted the foundations of electricity and magnetism, transforming the field from a largely descriptive subject into a quantitative discipline in which one could employ the calculus to tackle realistic engineering problems.

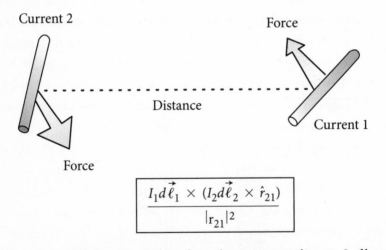

$$\frac{I_1 d\vec{\ell}_1 \times (I_2 d\vec{\ell}_2 \times \hat{r}_{21})}{|r_{21}|^2}$$

**Figure 4.3.** Ampère's Law: The magnetic forces that two current elements, $I_1$ of length $dl_1$ and $I_2$ of length $dl_2$, separated by a distance $r_{21}$, exert on each other. Each element can point in an arbitrary direction, and they can be separated by an arbitrary distance. Ampère's formula is shown boxed.

## A SYMPATHETIC UNDERSTANDING

Ampère titled his paper "Theory of Electrodynamic Phenomena, Uniquely Deduced from Experience." But as Henri Poincaré remarked in 1905 about "Ampère's immortal work," Ampère's laws could *not* have been deduced from experience, because he had no infinitesimal currents to experiment with. Only what Einstein called intuition or a "sympathetic understanding of experience" could have led him from observations of entire circuits to a law for infinitesimal current elements. In his encyclopedic *A Treatise on Electricity and Magnetism*, James Clerk Maxwell later wrote:

> The experimental investigation by which Ampère established the laws of the mechanical action between electric currents is one of the most brilliant achievements in science.
>
> The whole, theory and experiment, seems as if it had leaped full grown and full armed from the brain of the "Newton of Electricity." It is perfect in form and unassailable in accuracy, and it is summed up in a formula from which all the phenomena may be deduced, and which must always remain the cardinal formula of electrodynamics.
>
> The method of Ampère, however, though cast into an inductive form, does not allow us to trace the formation of the ideas which guided it. We can scarcely believe that Ampère really discovered the law of action by means of the experiments which he describes. We are led to suspect, what, indeed, he tells us himself, that he discovered the law by some process which he has not shown us, and that when he had afterwards built up a perfect demonstration, he removed all traces of the scaffolding by which he had built it.

Maxwell, who performed his own magic, recognized that miraculous discoveries seem to leap out of the mind's invisible Dirac sea, elicited by intuition.

## FARADAY: MOVING MAGNETS CREATE ELECTRIC CURRENTS

Back to the phenomena. Michael Faraday, impressed by Oersted's discovery that one current could push on another, became convinced that one current alone should be able to *produce* another current too. In 1831 he ran a *steady* current through one circuit and watched for a steady current in another. Disappointed at observing nothing, he switched off the current in the first circuit—and unexpectedly observed a flicker of current in the second. The *change of current* in the first wire induced an electric current in the second.

Because he knew that currents behave like magnets, Faraday conjectured and then demonstrated that moving a magnet near a closed loop of wire could induce an electric current in it too, as illustrated in Figure 4.4. Well acquainted with the force that magnets exert on

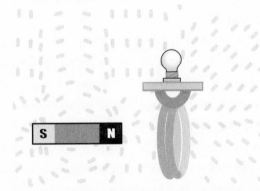

**Figure 4.4.** Faraday's discovery of induction: a moving magnet induces an electric current in a circuit.

currents, Faraday quickly figured out how to build the first electric motor by surrounding a loop of current-carrying wire with magnets, which made it rotate. Then he did the inverse: by rotating a loop of wire in the vicinity of a magnet he produced a current in the loop and created the first electric generator. The discovery of phenomena and the development of theory have played leapfrog in the history of electromagnetism, a fruitful progression missing in the development of models of financial markets.

## FARADAY IMAGINES FORCE-TRANSMITTING LINES

Until Faraday everyone assumed that charges and currents pushed on each other across empty space without any delay. All interactions were thought to reside purely within the objects that felt them. Faraday knew little mathematics beyond algebra, but he was adept at thinking spatially. In a Spinozan act of identification with the object, Faraday dreamed up *lines of force*. He pictured every charge exuding *electric lines of force* that flowed though space to create an invisible web, transmitting and exerting electric forces on charges in other locations. He visualized *magnetic lines of force* similarly pervading space to attract iron filings and to influence distant currents. Figure 4.5 shows the schematic lines of electric and magnetic forces necessary to account for the distant electrical influences of charges and magnets.

Faraday recast all the relationships between electric and magnetic forces known at that time in terms of equivalent dynamic relationships between the lines, so that the laws of electromagnetic interactions became laws of interacting lines. Thus, for example, his own discovery that "magnets in motion induce electric currents in circuits" was transformed into the more abstract proposition "A change in the number of magnetic lines of force creates electric lines of force."

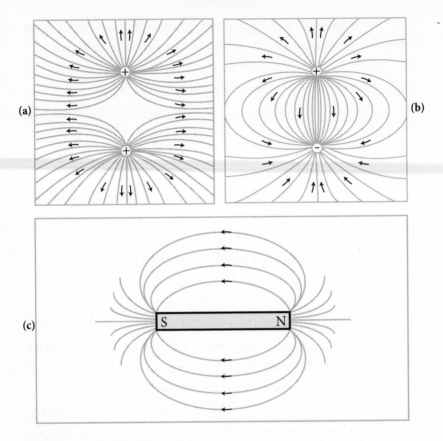

**Figure 4.5 (a)** Electric lines of force from two positive charges. **(b)** Electric lines of force from one positive and one negative charge. **(c)** Magnetic lines of force emanating from a bar magnet.

In a paper he published in 1846, almost 60 years ahead of its time, Faraday presciently began to regard electromagnetic radiation as existing in its own right, without the need for a medium to flow through:

The view which I am so bold to put forth considers, therefore, radiation as a kind of species of vibration in the lines of force which are known to connect particles and also masses of matter together. It endeavors to dismiss the aether, but not the vibration.

The lines originated as metaphor; Faraday began to liberate them; shortly thereafter, Maxwell set them free.

## MAXWELL MODELS THE LINES

> From a long view of the history of mankind—seen from, say, ten thousand years from now—there can be little doubt that the most significant event of the 19th century will be judged as Maxwell's discovery of the laws of electrodynamics. The American Civil War will pale into provincial insignificance in comparison with this important scientific event of the same decade.
>
> —Richard Feynman, *Lectures on Physics*

The historic achievement of James Clerk Maxwell, a Scottish mathematician and theoretical physicist with a very practical streak, was the unification of electricity and magnetism into a consistent set of equations for electromagnetic theory. Feynman, no mean achiever himself, is accurate about the magnitude of Maxwell's discovery, which was the midpoint on the trajectory from Newton's discovery of the laws of mechanics to Einstein's theories of relativity. Maxwell was a bird who flew from branch to branch and field to field, not only developing electromagnetic theory but also playing a major role in the creation of thermodynamics, statistical mechanics, and the theory of controlling engineering devices.[1]

Maxwell made his first assault on electromagnetism in 1856, in a paper entitled "On Faraday's Lines of Force." Faraday's imaginary electric and magnetic lines of force reminded Maxwell of the streamlines in the flow of a fluid. By the mid-1850s the theory of fluid flow, or hydrodynamics, was mathematically sophisticated and powerful. Maxwell, inspired by Faraday's vision, decided to model Faraday's lines of force as "the motion of an imaginary fluid," hoping to get some insight from what he thought of as only an analogy rather than reality. It was a warm-up exercise for Maxwell, who clearly understood the difference between a theory and a model. Working by analogy, he explicitly hoped to "avoid the dangers arising from a premature theory professing to explain the cause of the phenomena." Maxwell's tentative model for electromagnetism was based on the better-understood theory of fluid flow and was meant to suffice until "a mature theory, in which the physical facts will be physically explained, will be formed." The result was a set of differential equations for the interactions between the still imaginary fluid lines of electric and magnetic force filling space.

## MAXWELL REIFIES THE LINES

Until now Maxwell had merely translated Faraday's intuition about imaginary lines of force into the more formal language of differential equations, using analogies with fluid flow as an aid to thinking. Then he crossed the threshold: in 1861, in a paper entitled "On Physical Lines of Force," he began to regard the lines as genuine stresses in a space-filling ether. In Maxwell's hands, Coulomb's law, Ampère's laws, and Faraday's law all became propositions about the dynamics of lines of force. Here are two examples of how the laws of forces on visible objects were rewritten as laws of invisibly interacting *lines of force*:

1. Ampère's law of force specified how two current elements pushed against each other. It was a statement about observable effects. Rephrased by Maxwell it became:

*Ampère's law:* An electric current creates *magnetic* lines of force.

In Maxwell's view, one electric current element produces magnetic lines of force that permeate the space around it. These lines, like a magnet, push on the other current element.

2. Faraday's law of induction stated that the fluctuation of an electric current in one circuit induces a current in another. Rephrased by Maxwell, it stated:

*Faraday's law:* A change in the *magnetic* lines of force creates *electric* lines of force.

A fluctuating electric current in the first circuit produces a corresponding fluctuation in the number of magnetic lines of force that fill the space around it. This fluctuation in turn creates electric lines of force that impel the charges in the second circuit to flow; hence an induced current. Charges and current no longer act on each other directly. Now their actions are mediated by the lines.

## MAXWELL MODIFIES AMPÈRE'S EQUATIONS

There is a glaring asymmetry between Ampère's law and Faraday's law. Whereas Faraday's law states that electric lines of force are created by a change in magnetic lines, Ampère's law says that magnetic lines are created only by electric currents. Why shouldn't changes in electric lines similarly induce magnetic lines? Guided by intuition and a sense of symmetry, Maxwell added an entirely new law for the production of lines of force, a law obtained by switching the words *magnetic* and *electric* in Faraday's law in (2) above, transforming it into:

3. *Maxwell's addition to Ampère's law:* A change in the *electric* lines of force creates *magnetic* lines of force.

Experiment didn't demand this law, and historians of science still disagree about the impetus that led Maxwell to make the addition. But what matters most is that, without experimental evidence, he perceived the necessary existence of a phenomenon that hadn't yet been observed.

## MAXWELL'S THEORY: THE FIELD ITSELF

Since Maxwell's time, physical reality has been thought of as represented by continuous fields, and not capable of any mechanical interpretation. This change in the conception of reality is the most profound and the most fruitful that physics has experienced since the time of Newton.
—Albert Einstein, "Maxwell's Influence on
  the Development of the Conception of Physical
  Reality," in *James Clerk Maxwell: A Commemorative
  Volume 1831–1931* (1931), 71.

Why the extreme praise? Because Maxwell changed the way physicists do physics. He examined the equations obeyed by the visible world, saw a pattern with something missing, completed it, and deduced the existence of electromagnetic waves. The imposition of an apparently missing symmetry has become the classic modus operandi of theoretical physics. The discovery of the Dirac equation, the unveiling of quarks, and the elucidation of the Standard Model all proceeded in similar style.

In 1864 Maxwell published "A Dynamical Theory of the Electro-magnetic Field." In this paper he christened the lines of force *the field*, writing, "The electromagnetic field is that part of space which contains and surrounds bodies in electric or magnetic conditions." The lines of force, once an aid to thinking, became a theory.

I like the phrase "a dynamical theory." Maxwell's equations describe interactive movement, the shapes of the electric and magnetic fields as they twist and curl through space and influence each other. Like most things and ideas, the field is observable only indirectly, via its effects, but no one doubts its reality.

## MAXWELL'S EQUATIONS: THE FIELD'S GEOMETRY—CURLS AND DIVERGENCES

As shown in Figure 4.6, a field in general can diverge like a fountain or curl like a halo.[2] Maxwell's equations elegantly describe the causes of the electromagnetic field's curls and divergences, which can be understood pictorially.

I shall denote the electric field by the symbol $\vec{E}$ and the color black, and the magnetic field by the symbol $\vec{B}$ and the color gray. The arrow $\vec{\phantom{x}}$ above each field's symbol indicates that it is a vector, meaning that at each point in space the field has not only a magnitude but also a direction, like a force, or a fluid flowing. Maxwell's

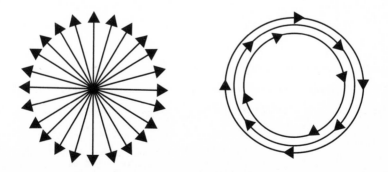

**Figure 4.6.** The configurations of fields: divergences and curls.

four famous equations concisely specify what causes $\vec{E}$ and $\vec{B}$ to diverge or curl, as illustrated in Figures 4.7–4.10.

To summarize, electric charges produce divergent $\vec{E}$ fields; electric currents produce curling $\vec{B}$ fields; time-varying $\vec{E}$ fields produce curling $\vec{B}$ fields; time-varying $\vec{B}$ fields produce curling $\vec{E}$ fields.

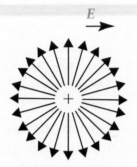

**Figure 4.7.** Maxwell's first equation: A positive electric charge is the source of a divergent electric field $\vec{E}$. Similarly a negative charge is the sink of a convergent $\vec{E}$ field. The number of electric lines of force is proportional to the charge.

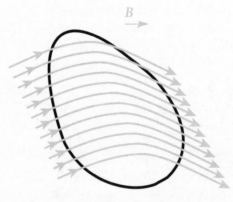

**Figure 4.8.** Maxwell's second equation: Magnetic fields $\vec{B}$ never diverge or converge, because there are no single magnetic poles found in nature to serve as their source.[3] Thus every line that flows *into* a closed surface must flow *out* again.

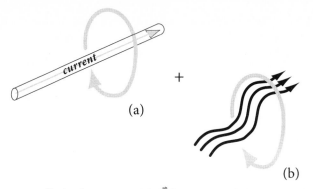

(a)

+

(b)

**Figure 4.9.** Maxwell's third equation: Since $\vec{B}$ lines cannot diverge, they can only curl. There are two independent ways to generate a curling magnetic field : (a) via a steady electric current (corresponding to Ampère's law), and (b) via a time-varying electric field (corresponding to Maxwell's modification of Ampère's law).

**Figure 4.10.** Maxwell's fourth equation: Electric fields can curl as well as diverge. Fluctuating magnetic fields produce curling electric fields (corresponding to Faraday's law).

## THE GREAT CONFIRMATION: LIGHT IS THE PROPAGATION OF ELECTROMAGNETIC WAVES

How did Maxwell know that his addition to Ampère's law was correct? Because that addition made it possible for waves of $\vec{E}$ and $\vec{B}$ fields to propagate indefinitely through empty space.

By combining the equations depicted in Figures 4.9a, 4.9b, and 4.10, you can see how waves can be produced and then propagate, as illustrated in Figure 4.11. It begins with an *oscillating* electric current, which, by Figure 4.9a, generates an *oscillating* curling magnetic field. The oscillation in time is crucial, because it is the magnetic field's oscil-

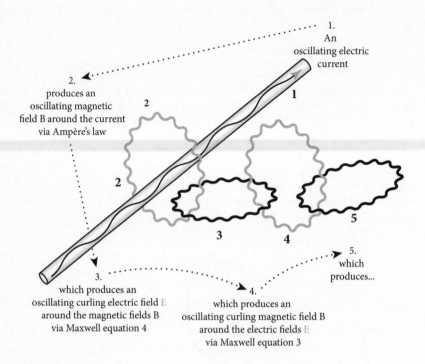

**Figure 4.11.** The generation and propagation of electromagnetic waves via Maxwell's equations.

Step 1: An oscillating electric current in the wire (an antenna)

Step 2: Produces an oscillating magnetic field via Ampère's law;

Step 3: Which, via Maxwell equation 4, produces a **curling oscillating electric field.**

Step 4: The extra term in Maxwell's equation 3 guarantees that the **oscillating electric field** will induce a *fresh* oscillating curling magnetic field.

Step 5: Which produces a **curling oscillating ...**

The diagram is meant to be schematic. Only parts of the field are shown. The wiggles in the loops are meant to indicate the oscillatory nature of the fields.

lations, according to Figure 4.10, that induce a similarly oscillating curling electric field, which in turn, according to Figure 4.9b, induces another oscillating curling magnetic field, which . . . This is, precisely, a wave! Waves of electric and magnetic fields propagate through space by leapfrogging, an oscillating $\vec{E}$ leading to an oscillating $\vec{B}$ leading to an oscillating $\vec{E}$ ad infinitum. You can think of the current-bearing wire in the center of the figure as an antenna that initiates the radiation, which then continues to re-create itself recursively.

Working out the mathematics of the leapfrogging, Maxwell discovered that the theoretical speed with which the waves propagated through space was about 186,000 miles per second, the measured speed of light. Therefore, he deduced, the light we see is in fact an electromagnetic wave whose frequencies happen to be visible to our eyes.[4] Heat, ultraviolet rays, radio waves, and X-rays are merely the same Maxwellian fields, differing from each other only in their wavelength. Without Maxwell's addition to the equations, there would have been no leapfrogging and no waves in the theory at all.

One can learn even more about the waves of light by examining Figure 4.12. There the black and gray loops give rise to each other sequentially, like a father and child building a tower of hands. Each $\vec{E}$ wave curls about the previous $\vec{B}$ wave, and each $\vec{B}$ wave curls about the previous $\vec{E}$ wave, so that at any point in space the electric and magnetic fields curl perpendicular to each other, and also perpendicular to the direction in which they move. That is, the waves propagate transversely, as illustrated. In this way Maxwell's equations explained the fact that light can be polarized. The transverse propagation of light is quite different from the so-called longitudinal propagation of sound waves, in which the variations in air pressure occur along the direction in which the wave travels.

In 1887 Heinrich Hertz used an antenna to create, transmit, receive, and reflect electromagnetic waves in his laboratory, entirely confirming the theory. About Maxwell's equations he later said:

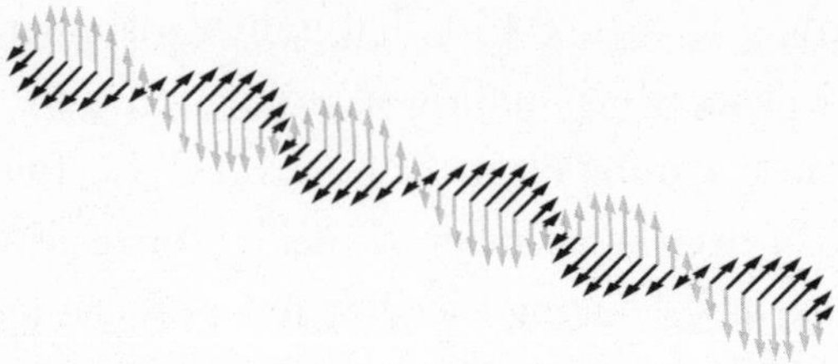

**Figure 4.12.** The transverse propagation of electromagnetic waves.

One cannot escape the feeling that these mathematical formulae have an independent existence and an intelligence of their own, that they are wiser than we are, wiser even than their discoverers, that we get more out of them than was originally put into them.[5]

## REALITY = PERFECTION; FACT = THEORY

There is no better or truer description of the physical phenomenon of light than to say it *is* precisely the electromagnetic field defined by Maxwell's equations. The electromagnetic field is not *like* Maxwell's equations; it *is* Maxwell's equations. While a model builder knows that his model airplane is *like* a true airplane,[6] and a climate modeler is aware that his equations only simulate the atmosphere, a physicist knows that light and Maxwell's equations are one and the same.

I see the trajectory of the discoveries of electromagnetism as follows:

From phenomena as facts
to lines of force as an aide to visualization
to analogies with fluid flow
to a theory of fields that has again become fact.

We advance by mounting new theories atop previous facts. If those theories prove correct, they become facts too.

## THE BEASTS OF THE FIELD

Who teaches us more than the beasts of the field and makes us wiser than the birds of the heavens?

—Job 35:11

The advances of the sublime theory of electromagnetism didn't end in the late nineteenth century. In 1897 J. J. Thomson discovered the electron, the minuscule carrier of the electric current. In 1900 Max Planck noticed the first hint of the quantum nature of light, and in short order the completeness of classical physics—Newton for matter and Maxwell for light—began to unravel. Einstein showed that smooth light waves consist of chunky quanta, and Schrödinger that chunky electrons satisfy smooth quantum mechanical wave equations.

The *idea* of the electromagnetic field is Maxwell's equations. The *idea* of the electron is the Dirac equation. Combining them both we obtain QED, the *idea* of electrons interacting with light quanta, the theory of relativistic *quantum electrodynamics*. The equations that compose it, Maxwell's and Dirac's, were complete by the late 1920s.

QED seemed to work magnificently, the result of calculations using the theory agreeing perfectly with experiment. But in the late 1940s atomic physicists, among them Polykarp Kusch and Willis Lamb at Columbia, discovered tiny but very precise anomalies in the radiation emitted by the single electron orbiting the proton in a hydrogen atom.[7] The wavelengths they measured didn't quite match those predicted by the straightforward use of QED. Theorists set about trying to explain the size of the discrepancies and were led to consider the possibility that they might stem from the interactions of the atom with the invisible Dirac sea itself.

From a theoretical point of view, what we call a single isolated electron is neither single nor isolated: it travels through the Dirac sea, which is filled to the brim with an infinite number of invisible negative-energy electrons. When you accept the existence of the sea as gospel rather than metaphor, the Dirac theory of a supposedly single electron is actually a theory of *many* particles. There *is* no solitary electron; the sea of electrons is the medium in which all particles live. Electrons can pop out of it and leave behind empty positron-holes into which other electrons can then sink in temporary oblivion.

This medium influences everything. The simple Coulomb attraction between a hydrogen atom's proton and its orbiting electron is altered by their immersion in the sea. The sea's invisible electrons repel the orbiting electron; the repulsion modifies the orbit and hence alters the frequencies of the light waves it emits when it transitions from one orbit to another.[8] Keeping careful track of this complexity—the interactions of the electron with the vast sea and with its own electromagnetic field—is immensely difficult, both conceptually and technically.

### Conceptual Difficulties: The Stand-Alone Electron Is Bare, the Real One Dressed

The real electron we see is not the stand-alone electron the Dirac equation begins with, undisturbed and private. Physicists call that hypothetical stand-alone electron the *bare* electron. The real electron, the only electron we can experiment on, suffers the drag of the sea; physicists call the real electron the *dressed* electron, because it is clothed by the bits of the sea that stick to it. The idea of *dressing* is a theoretical construct to take us from the imperfect unreal stand-alone electron to the perfect real one that exists.

Because it is clothed, the dressed electron moves differently than it would if there were no sea. To accurately calculate the properties of a real electron in a real hydrogen atom, one must deal with the motion of a dressed electron.

## Technical Difficulties: Calculations Produce Infinities . . .

The drag of the sea changes the electron's resistance to motion, not unlike the way a car feels more massive if you drive it with the hand brake on. Because of the sea, the mass of the dressed electron (colloquially called the "dressed mass") differs from the mass of the bare electron (the "bare mass").

Similarly the perceived charge of a real electron, the dressed charge, differs from its bare charge because the negative-energy electrons in the sea surround and shield the bare electron, modifying the charge it displays to the outside world. The dressed electron is a bare electron immersed in a medium. Just as light travels more slowly through glass, so the electron behaves a little differently as it moves through the sea.

You would think that these sea changes should be small, because electromagnetic forces in general are relatively weak. But you can calculate their size in the full theory of QED that takes account of the medium, and—*the horror!*—they turn out to be infinite. The numerical difference between the dressed mass and the bare mass is infinity. Equally unfortunate and equally infinite is the difference between the dressed charge and the bare charge. These are the true beasts of the field.

## . . . But Physicists Figure Out How to Evade Them

Formal mathematics can help physics, but has never been allowed to hinder it. New developments in mathematics, from calculus to topology, have often been initiated by physicists who, by means of intuition and persistence, have sneakily but sloppily invented new kinds of mathematics that were only later made rigorous by purists. Newton invented the calculus in the seventeenth century to handle mechanics, and its foundations were satisfactorily cleaned up years later by Augustin-Louis Cauchy and his contemporaries. In the late 1940s, reconnoitering around the technical difficulties of the Dirac sea, Richard Feynman, Julian Schwinger, and Shin'ichirō Tomonaga found an ingenious way to suppress the technical infinities of quantum electrodynamics by means of a judicious combination of extreme care and chicanery. Their starting point was never to forget that the *normal* quotidian electron we "see" every day is not the bare electron. The normal electron is the *fully dressed electron*. Therefore, when we look at an electron, with our eyes or our apparatus, what we see—what we know as fact—is its dressed mass and charge.

Consequently there is no point in using the theory to calculate corrections to the value of the mass and charge, since including them would be the *double counting* of an effect that is already there. But that doesn't mean that the effects of the sea don't matter. They do, *but not for the purposes of calculating the mass and charge of the electron,* since these we already know from measurement. We should instead use the theory only to calculate the *incremental* effects of the sea on the motion of an electron in the hydrogen atom, over and above its effects on the mass and charge.

**Now for the Chicanery**

The fact is that we don't care what the Dirac sea does to free electrons, because we have already accounted for the drag of the sea on the free electron by making use of the observed values of its mass and charge. All so-called free electrons are already conceptually dressed by the sea. Given that fact, what is important is the drag on the bound electron *over and above* the drag on the free electron. That's the effect we can actually observe.

What Feynman, Schwinger, and Tomonaga noticed, mirabile dictu, was that though the drag on both bound and free electrons was infinite, the *difference* between them was finite! All the infinities can be dodged if we calculate only the difference between the drag on the free and the drag on the bound electron. And that is the only difference that is observable. If differences are all we calculate, then nothing in any QED calculation is infinite.

## The New Normal

Physicists call this process *renormalization,* a recalibrating of normality to accommodate the recognition that the "normal" reality and perfection of the world correspond not to what goes *into* the theory, but only to what comes *out* of it when you have solved it. The normal isolated free electron we see is the electron dressed by the sea, not the bare stand-alone Dirac electron. One must redefine "normal" to be the final reality, not the original imperfection.

Feynman, Schwinger, and Tomonaga painstakingly calculated the effects of the Dirac sea on the orbiting electron inside a hydrogen atom relative to the new normal. They found that the sea produced minute changes in the frequencies of light waves emitted by hydrogen, and that the calculations agreed with then contemporary experiments to the stunning accuracy of one part in several thousand. Nowadays the theory's predictions are accurate to one part in 100 billion or so, or, as Feynman put it, equivalent to measuring the distance from San Francisco to New York with the accuracy of a hairbreadth. Renormalization works unimaginably well.

The pragmatic and cunning lesson it teaches is this: use your theories to calculate only what you can really measure, and carefully avert your eyes from everything else.

> Nature will reveal nothing under torture; its frank answer to an
> honest question is "Yes! Yes!—No! No!"
>
> —Goethe, *Maxims and Reflections*

Financial modelers use a process similar to renormalization to force their less than perfect, less than real models to fit the world they observe. They call this process *calibration,* the tuning of parameters in a model until it agrees with the observable prices of liquid (i.e., easily tradable) securities whose values we know. Only when a model is forced to be consistent with this "normal" state of markets can we reasonably use it to calculate the value of "abnormal" securities whose prices we don't know. But calibration in finance works much less well than renormalization in physics: in physics the normal and abnormal are governed by the same laws, whereas in markets the normal is normal only while people behave conventionally. In crises the behavior of people changes and normal models fail. While quantum electrodynamics is a genuine theory of all reality, financial models are only mediocre metaphors for a part of it.

## ELECTROMAGNETISM AS METAPHOR

Like all underlying facts and theories, electromagnetism serves as a metaphor with which to understand ourselves and the world around us, as pointed out by Owen Barfield in his book *History in English Words*:

> The phrase "high tension," used of the relation between human
> beings, is a metaphor taken from the condition of the space between
> two electrically charged bodies. At present many people who use such
> a phrase are still half-aware of its full meaning, but many years hence
> everybody may be using it to describe their quarrels and their nerves

without dreaming that it conceals an electrical metaphor—just as we ourselves speak of a man's "disposition" without at all knowing that the reference is to astrology. . . .

The scientists who discovered the forces of electricity actually made it possible for the human beings who came after them to have a slightly different idea, a slightly fuller consciousness of their relationship with one another. They made it possible for them to speak of the "high tension" between them. So that the discovery of electricity, besides introducing several new words (e.g., electricity itself) into our everyday vocabulary, has altered or added to the meaning of many older words, such as battery, broadcast, button, conductor, current, force, magnet, potential, tension, terminal, wire, and many others.

## QUANTUM DREAMS

Once, after long and sustained efforts with the theory of quantum mechanics in graduate school, I had an exhausting but exhilarating dream: my entire body was a Schrödinger wave that *had to* satisfy Schrödinger's famous wave equation. In my dream I struggled to flex my body so as to make it undulate like a three-dimensional violin string, contorting myself to ensure that I satisfied the boundary conditions of the differential equation, oscillating inside the box I dream-inhabited, but pinned down at its walls.

I like to imagine that it was by some similar visceral but deeper intuition that the discoverers of electromagnetic theory and QED merged with their subject.

# EPILOGUE

Everything that we call Invention or Discovery in the higher sense of the word is the serious exercise and activity of an original feeling for truth, which, after a long course of silent cultivation, suddenly flashes out into fruitful knowledge. It is a revelation working from within on the outer world, and lets a man feel that he is made in the image of God. It is a synthesis of World and Mind, giving the most blessed assurance of the eternal harmony of things.

—Goethe, *Maxims and Reflections*

# III. MODELS BEHAVING BADLY

# CHAPTER 5

# THE INADEQUATE

*Financial models are not the physics of markets • Method without content leads to sophistry • Finance is about expectations • Uncertainty is everywhere • The Efficient Market Model assumes that all uncertainty is quantifiable • The only reliable principle: The Law of One Price • Return is proportional to risk • The pleasure premium • The unreliability of CAPM • Why the Efficient Market Model fails • The unbearable futility of modeling*

There is nothing so terrible as activity without insight.
—Goethe, *Maxims and Reflections*

I began my professional life as a physicist, studying fundamentals and mastering theory. Then, in 1985, I migrated to the center of the quant world at Goldman Sachs. My colleagues were as smart as academics but more interesting. The work was an interdisciplinary mix of modeling, mathematics, statistics, and programming, all aimed at trying to value securities for trading desks. Quants were the theorists, traders were the experimentalists, and we collaborated to develop and explore our models. Though the aim was moneymaking, the environment was collegial and the techniques were remarkably similar to those I'd used in physics. Within a few months I had met and begun to collaborate with Fischer Black, the coinventor of the Black-Scholes option pricing model, which regards markets as equilibrium-seeking systems and models them by analogy with the physics of heat diffusion.

I plunged into work, reading and modeling and programming. I learned the elegant logic behind Black's work and witnessed the depth of his thinking. I worked closely with traders who survived and prospered by using models. Soon I began to believe it was possible to apply the methods of physics successfully to economics and finance, perhaps even to build a grand unified theory of securities.

After twenty years on Wall Street I'm a disbeliever. The similarity of physics and finance lies more in their syntax than their semantics. In physics you're playing against God, and He doesn't change His laws very often. In finance you're playing against God's creatures, agents who value assets based on their ephemeral opinions. The truth, therefore, is that there is no grand unified *theory* of everything in finance. There are only *models* of specific things. I say "model" because finance relies on modeling the mental qualities of stocks and markets as though they satisfy the theories of the physical world. The analogies, though not unfounded, are partial and flawed. That doesn't mean that modeling in finance is a waste of time; it means that you have to understand what models are best used for, and then be very careful not to discard your common sense, as I shall explain in the next chapter. In this chapter I want to describe the basic principles of modern financial models and highlight the use of metaphors and analogies as they appear. I want to emphasize the assumptions being made and comment on their plausibility.

If I were to summarize this chapter in one sentence, I would write: *Financial modeling is not the physics of markets.*

## FINANCE IS NOT MATHEMATICS

Recently a visiting professor of finance gave a seminar in my department at Columbia. Later, talking about his work in my office, he justified a result in his presentation by invoking what he referred to as "the fundamental theorem of finance."

Isn't it strange, I thought afterward, that finance has a fundamental theorem (and that, despite more than twenty years in the field, I wasn't quite sure what it was)? Why doesn't physics, with its superior precision, have a fundamental theorem too?

After he left I Googled "the fundamental theorem of physics" and found nothing at all. So I Googled just the phrase "fundamental theorem" and soon found *the fundamental theorem of arithmetic*: "Every natural number greater than 1 can be written as a unique product of prime numbers." That isn't hard to understand, and you can see why someone might call it fundamental. I Googled further and found *the fundamental theorem of algebra*: "Every polynomial equation of degree *n* with complex number coefficients has *n* complex roots." This is harder to understand, because algebra is more difficult than arithmetic, but still it's fairly straightforward.

Then I Googled "the fundamental theorem of finance." I didn't have to look too far down the list of URLs to discover the following statement:

**Fundamental Theorem of Finance.** Security prices exclude arbitrage if and only if there exists a strictly positive value functional, under the technical restrictions that the space of portfolios and the space of contingent claims are locally convex topological vector spaces and the positive cone of the space of contingent claims is compactly generated, that is, there exists a compact set $K$ of $X$ (not containing the null element of $X$) such that

$$C = \{x \in X : x \geq 0\} = \bigcup_{\lambda \geq 0} \lambda K$$

How can anyone think that this incomprehensible statement is fundamental to finance, a field whose focus is the management of money and assets? Why should finance possess a theorem and physics lack one?

## LAWS ARE NOT THEOREMS

Physics is concerned with the world about us. The world may or may not have laws, but it cannot have theorems.

### The Theorems of Mathematics

Theorems are if-then relations. The *ifs* are conditions; the *thens* are the consequences that follow. In geometry Euclid's axioms and postulates are the unquestioned *ifs*, more or less apparently self-evident properties of points and lines. One postulate, for example, is that it is always possible to draw a straight line between any two points. One axiom is that things equal to the same thing are equal to each other. Difficult, though not impossible, to argue with.

Euclid's points and lines are abstracted from those of nature. When you get familiar enough with the abstractions, they become tangible. Even more esoteric abstractions, such as the infinite-dimensional Hilbert spaces that form the mathematical basis of quantum mechanics, seem real and imaginable to mathematicians. Nevertheless the theorems of mathematics are relations between the abstractions, not between the realities that inspired them.

## The Laws of Physics

Science, in contrast, has laws. Laws are not contingent. They describe the way the universe works, unconditionally. Newton allows us to guide rockets to the moon. Maxwell enables the construction of radios and TV. Thermodynamics makes possible the construction of combustion engines, which convert heat into mechanical energy. QED drives all our electronic gadgets and networks.

Like mathematics, physics deals with abstractions, but there is a unity between the abstraction and the object it represents. The laws we discover are valid relations between one abstraction and another, but they are also valid for the objects "beneath" the abstractions, providing a connection between the Idea mode and the Extension mode. Hayek summarized this perceptively when he wrote:

> The task of the physical sciences is to replace that classification of events which our senses perform but which proves inadequate to describe the regularities in these events, by a classification which will put us in a better position to do so.[1]

# FINANCE MISUNDERSTOOD

> Content without method leads to fantasy; method without content to empty sophistry.
>
> —Goethe, *Maxims and Reflections*

Finance's objects of interest—markets, money, assets, securities—are also abstractions. The aim of finance, like that of physics, is to find not only the relationships between the abstractions themselves, but also the relationships between the realities they represent. Finance cannot be a branch of mathematics, and therefore should not have a fundamental theorem. Only those who don't understand the difference between the contingent and the actual, between their inside and their outside, can imagine theorems of finance. If you cannot distinguish between God's creations and man's idols, you may mistake your models for laws. Unfortunately many economists are such people. If you open up the prestigious *Journal of Finance,* one of the select number of journals in which finance professors must publish in order to get tenure, many of the papers resemble those in a mathematics journal. Replete with axioms, theorems, and lemmas, they have a degree of rigor that is inversely proportional to their minimal usefulness.

Physicists, brought up on a diet of astounding theories and successful models, have the ability to distinguish a theory from a model and a good model from a bad one. Economists for the most part have never seen a genuine theory, and so discrimination is harder. The simple models they work with fail to reflect the complex reality of the world around them. That lack of success is not the fault of economists, for people have proved difficult to theorize about, and we still await an understanding of Spinoza's adequate causes for their behavior. But it is the economists' fault that they take their simple models so seriously.

Finding the truth about nature takes cunning and intuition. The invisible worm of financial economics is its dark secret love of mathematical elegance regardless of its efficacy, and its belief that rigor can replace fact and intuition.

## SECURITIES AND MARKETS

> It would be most desirable . . . to base the language for the
> details of a particular area on the area itself.
> —Goethe, *Maxims and Reflections*

Understanding the principles of financial models requires a familiarity with the objects of interest.

### SECURITIES

A *security* is a promise by its seller (a person or an institution) to make *future* payments to its buyer. A security represents *present* financial value, and financial theory aims at figuring out the magnitude of that value—what the security is worth—and why.

### Debt Securities

The simplest kind of security is a personal loan: you lend money to someone who signs an IOU to repay you. You can regard the IOU as a debt security you bought from the borrower by paying him the amount you lent him. A Treasury bond is a debt security too: you lend a sum of money to the U.S. government and it promises to repay the principal after 30 years and, during the interim, to make interest payments to you, equal, for example, to 4.25% per annum.

Corporations can borrow money by selling IOUs called corporate bonds. Conversely, if you hold a mortgage on your house, you sold an IOU to the bank that requires you to pay it future interest and principal, with the house as collateral in case you break your promise.

Debt securities are characterized by precise specifications, and their future payoff is usually spelled out very clearly.

## Equity Securities

Equity securities have less certain payoffs. The most common equity security is a share of stock, an investment in a company that grants you a stake in its business and consequently entitles you to a proportion of the profits. The invention of publicly owned companies and the limited liability of their owners for losses was one of the great facilitators of the spread of capitalism, and the inability or unwillingness to prosecute an individual for a corporation's misdeeds is no doubt also a facilitator of corporate irresponsibility.

When you buy a share of stock, the company can use your payment to run its business exactly as it pleases.[2] Instead of a promised rate of interest, you get your share of any future earnings. Because the existence and quantity of future earnings are uncertain, there is an implicitly contingent quality to the stock, which is therefore more risky than a bond.

Most shares of stock also give shareholders the legal right to vote at annual meetings of the corporation and to have a say in selecting boards and making decisions about potential mergers. Although these rights are not always exercised and are hard to quantify, they are valuable. This additional layer of strategic complexity is part of the reason that stock markets get more public attention than debt markets; everyone loves a good story.

## Derivatives

Futures contracts, forward contracts, and options on stock are securities that involve even more uncertainty than equity. They are *derivative* securities; the seller of any one of these securities is obliged to make future payments to the buyer that depend on the future stock price, so that the securities *derive* their value from the stock. *Derivative contracts* are explicitly contingent and are often referred to more pedantically as *contingent claims*.

For Spinoza, *love* is a derivative of *pleasure*. Analogously, a *call option* derives its value from the underlying *stock* because, at the expiration of the option, the seller must pay the buyer the amount by which the stock price has risen since the purchase of the option. If the stock price has fallen, the seller owes nothing.

A call option is asymmetric, paying off when the stock price rises but not when it falls. A forward contract is symmetric: if the underlying stock price rises, the buyer of the contract gets the *pleasure* of receiving from the seller a payment corresponding to the increase in price; if, however, the stock price falls, the buyer suffers the corresponding *pain* of having to reimburse the seller for the decrease.

Because of their contingent nature, the values of derivative securities can vary dramatically with the price of the stock. Extreme sensitivity is always dangerous, especially so when attempts to calculate the sensitivity depend on an inaccurate model. This was the case with collateralized default obligations (CDOs). Their sensitivity to housing prices played a major role in initiating the financial crisis that began in 2007. Their values depended keenly, but opaquely, through several levels of derivatives, on the value of the underlying housing market. When the housing market crashed, the CDOs spread that market's weakness through the entire investment world to the large numbers of institutions that had bought them.

## MARKETS

Securities are traded on markets, physical or electronic locations where participants meet to exchange securities for money. Markets originated in a physical marketplace where people went from stall to stall and shouted bids and offers to each other in a crowd. Nowadays financial markets are increasingly situated on farms of computer servers distributed around the world.

The most straightforward are *listed* markets, which post in some central location their participants' offers to buy or sell a security, as well as a record of its price at the most recent sale. Listed markets deal with popular standardized securities that are easily described and available in quantity—stocks, for example, or fairly simple, so-called vanilla options. Less uniform securities that come in many similar but not identical variations—bonds, for example, which can have an extensive variety of maturities and coupons—commonly trade in *over-the-counter* markets, where buyers and sellers negotiate prices either face-to-face, by telephone, or electronically. On a listed market everyone can see the most recent prices, so one is less likely to be fleeced because of ignorance. Nevertheless many securities that should trade on listed markets still trade over the counter because the heavyweights in a market prefer to maintain the advantage that opaque pricing gives them over smaller participants.

News and information affect prices, which themselves are news and information. The history of prices and trading volumes from listed markets is therefore critical in developing financial models. Many agents in financial markets try to leapfrog the information flow by means legal (using algorithms to try to predict future price movements from past ones) or illegal (paying for *insider* information, a term whose precise scope keeps expanding to include formerly common behavior but not fast enough to keep up with people's ability to stay one step ahead).

## PRICE, VALUE, UNCERTAINTY

Price is what it costs to buy a security. Value is what you think it is worth. The difference between the two is what modeling and investing are about.

Worth is a mental quality, a matter of opinion and therefore subject to uncertainty. In contrast to the value of the electron's mass or charge, there is nothing absolute about the value of a financial asset.

In physics you can travel a very long way before you run into uncertainty. Newton's laws of motion and gravitation, Boyle's law relating the pressure and volume of ideal gases, Maxwell's equations for electromagnetism, and the study of the transformation of heat into mechanical energy in thermodynamics—all are triumphs of deterministic understanding. You begin to encounter the theory of randomness only when you try to understand the link between invisible atoms and the familiar macroscopic world.

In finance the thread of uncertainty emerges from the start. We cannot know how the value of a security will change through time because we don't know how the future will affect the promises made by its sellers. Value is determined by people, and people change their minds.

### Quoting Value

You can quote value in any currency, such as U.S. dollars, British pounds, or euros. Money is just another asset. What differentiates money from other securities is that, in the short run,[3] it's the least risky financial asset and is therefore useful as a medium of exchange and a common denominator for valuing other assets. Also, it's the asset the government decrees that you use to pay your taxes, and this gives it a very privileged status.

Nevertheless you don't have to quote value in terms of dollars or euros; you can quote value in ounces of gold or barrels of oil or shares of Google stock or sticks of Wrigley gum. If you are estimating the value of an oil company, for example, you might be better off valuing it in terms of barrels of oil rather than dollars, and letting the price of a barrel of oil take care of the relationship to dollars. In valuing a security, a good start is to think about the best currency to use for the model.

## THE EFFICIENT MARKET MODEL

The great financial crisis of 2007–2008 triggered an intense interest in the nature of quantitative financial models and their apparent inability to predict the disasters that occurred. The temptation to blame complex mortgage models for the market's near-dissolution was initially overwhelming. With time, observers have become more measured; Paul Krugman and Robin Wells have pointed out that identical disasters occurred even in markets where only simple mortgages (which required no sophisticated models) were traded— Spain, for example.[4]

At the core of modern finance is the Efficient Market Model, a powerful and simplistic model of human behavior. Despite its problems, it is worth a careful examination. To expose its assumptions and analogies is the purpose of this chapter.

### A Share of Stock

Because stocks are among the most common securities, I will use a share of stock as a typical risky security. Other securities—bonds, currencies, commodities, mortgages, real estate, and so forth—are also risky, but a share of stock will illustrate the idea adequately.

A company—take Apple, for example—is a tremendously complex and structured endeavor. Apple has tens of thousands of employees, owns or leases buildings in many countries, designs products ranging from desktop and laptop computers to iPhones and iPads to power plugs and cables, manufactures some of these on its own, outsources the manufacturing of others to China, distributes its products through its website and stores and also via third parties, and sells music and videos over the Internet. Apple advertises, provides product support, maintains websites, and carries out research and development.

The entire economic value of the organization is reflected in one number: the listed price of a share of its stock. This price is the amount of money most recently required to buy or sell just *one* incremental share of the company.[5] Financial modeling is an attempt to project the value of the entire enterprise into that number. It aims to tell you what you should pay today for a share of the company's, and so the stock's, future performance.

If your model indicates that the stock is worth more than its most recent price, you can buy it and hope to make a profit. For centuries, therefore, men and women have tried to predict the magnitude and direction of changes in the price of a bushel of wheat, an ounce of gold, or a share of stock, from moment to moment, day to day, month to month, or year to year. They haven't had any overwhelming success. Some predictors work from fundamentals, dissecting an entire company, its management, products, pipeline, budget, and style, as well as the state of the economy in which it operates. Others rely on technical analysis, a combination of rational and magical thinking that involves spotting the repetition of patterns in the trajectory of stock prices. None of these models works well consistently.

## Jujitsu Finance

It's a fact, then, that no one is very good at predicting stock prices. Faced with this failure, a school of academics associated with Eugene Fama at the University of Chicago in the 1960s developed what has become known as the Efficient Market Hypothesis, which I prefer to call the Efficient Market Model (EMM), since it's a model of a hypothetical world rather than a correct hypothesis about the one we inhabit.

I was a persevering student of physics when the EMM became popular, though I knew nothing of it. Over the years many formulations have evolved, some more formal and rigorous, some less so. But if you take care not to get carried away, as many faux-precision-worshipping economists do, if you can avoid the temptation to define *strong, weak,* and other kinds of "efficiency" as though you were dealing with a mathematical system rather than the world of humans and markets, then you will recognize that the EMM acknowledges the following true fact of life:

> It is well-nigh impossible to successfully and consistently predict what's going to happen to the stock market tomorrow based on all the information you have today.

The EMM formalizes this experience by stating that it is impossible to beat the market systematically because the current prices of stocks reflect all the information we have about the economy and the market. Only new information can cause the prices to change.

Converting their failed attempts at systematic stock price prediction into a fundamental postulate of their field was a fiendishly clever jujitsu response on the part of economists. It was an attempt to turn weakness into strength: "I can't figure out how things work, so I'll

make the inability to do that a principle on which to base a theory." The EMM, beneath its formal cloak, is simply an assumption about human behavior. It's therefore either right or wrong.

### In Efficient Markets, Price Equals Value

It's hard to argue with the statement that markets are unpredictable. The EMM goes further, though, and claims that at any instant current prices reflect all current and past information, and that therefore the best estimate of value is the current price. Stated that way, it sounds more dubious. Anyone with hindsight can see that the market is sometimes wrong about value.

Fischer Black, the codiscoverer of the Black-Scholes Model for valuing options, which grew out of the EMM, was a brilliant and original financial theorist, but also a determined realist. In his widely read paper "Noise" he acknowledged, as few academics do, the impossibility of assigning a precisely accurate value to a security:

> All estimates of value are noisy, so we can never know how far away price is from value.
>
> However, we might define an efficient market as one in which price is within a factor of 2 of value, i.e., the price is more than half of value and less than twice value. The factor of 2 is arbitrary, of course. Intuitively, though, it seems reasonable to me, in the light of sources of uncertainty about value and the strength of the forces tending to cause price to return to value. By this definition, I think almost all markets are efficient almost all of the time. "Almost all" means at least 90%.

Even this estimate of the discrepancy between price and value may be optimistic. No one knows how to truly determine value or

even exactly how to define it, except by means of a model, and so the discrepancy between price and value can never be determined.

The scientific development of financial modeling is bound to be shaky when it starts from an inability to define value clearly.

## UNCERTAINTY VERSUS RISK

Present value is ill-defined and future value is uncertain. Uncertainty implies risk; risk means danger; danger means the possibility of loss. In economics, thoughtful people have come to distinguish between *quantifiable* and *unquantifiable* uncertainty.[6]

### Unquantifiable Uncertainty

Unquantifiable uncertainty is, for example, the likelihood of a revolution in China or the detonation by terrorists of a nuclear bomb in midtown Manhattan. These events are unlikely, but there is no reliable method of estimating their odds. When someone says there is a one-in-a-million chance of a terrorist attack, it's an unverifiable guess rather than a probability. Though it sounds more reasonable to calculate the probability of a large earthquake, a purely physical phenomenon, to do so requires a model of the Earth's crust and its motion. No one knows how accurate such models are, and hence there is no way of truly estimating probabilities based on them. The best you can do with unquantifiable uncertainty is to be aware of it and aware of your inability to quantify it, and then to act accordingly. In trying to make sense of specifying the odds of a large earthquake, Freedman and Stark argue that common sense is best:

> Another large earthquake in the San Francisco Bay Area is inevitable, and imminent in geologic time. Probabilities are a distraction. Instead

of making forecasts, the USGS could help to improve building codes and to plan the government's response to the next large earthquake. Bay Area residents should take reasonable precautions, including bracing and bolting their homes as well as securing water heaters, bookcases, and other heavy objects. They should keep first aid supplies, water, and food on hand. They should largely ignore the USGS probability forecast.[7]

A similar approach to financial models would be wise.

## Quantifiable Uncertainty

In rare and contrived circumstances, such as card games and roulette, uncertainty is (almost) quantifiable, in which case F. H. Knight calls it *risk*. A classic example is the uncertainty involved in tossing a coin: Will it come up heads or tails? For an unbiased coin, the probability of heads or tails is $1/2$, which means that if you toss the coin enough times, the ratio of heads to tails will approach unity ever more closely as the number of coin tosses increases. Similarly one can determine the probability for three successive heads followed by two successive tails to be $(1/2)^5$, or $1/32$. These are so-called frequentist probabilities, an interpretation of the notion of probability in terms of the relative frequency of a particular outcome in a large number of identical trials.

For a Platonic coin, the uncertainty is quantifiable. There is, however, no realizable Platonic coin, just as there is no real-world Euclidean point or line. Any actual coin is inevitably biased, however slightly. Furthermore it experiences variable forces with each successive flick of the thumb, gusts of moving air as it spins, irregularities of the floor as it bounces. The chain of these events during the trajectory of the coin makes every detail of the environment so important that, counterintuitively, individual details become unimportant. If everything affects you so greatly, then nothing affects you very significantly. Consequently the coin is effectively uncoupled from the environment. Under those circumstances frequentist probabilities work well and uncertainty becomes risk.

For frequentist probabilities to be meaningful, one must be able to subject a system to repeated independent identical circumstances. You can do this with coins because their history is unimportant. In human affairs, history matters, and people are altered by every experience. To take a topical example, credit markets after the great financial crisis of 2007–2008 did not behave as credit markets did before, despite an identical lowering of interest rates, because lenders and borrowers were unforgettably affected by their experiences of the crisis.

It's not only the past that leaves its trace on humans. In physics, effects propagate only forward through time, and the future cannot affect the present. In the social sciences the imagined future can affect the present, and thereby the actual future too.

Despite this, the Efficient Market Model stubbornly assumes that all uncertainties about the future are quantifiable. That's why it is a model of a possible world rather than a theory about the one we live in.

## RISK DEMANDS A POSSIBLE REWARD

To buy a stock is to take a risk, voluntarily. One does so only if the (imagined) possible rewards outweigh the (imagined) risk of loss.

Physics models begin with the *current* state of the world and evolve it into the *future*. Financial models begin with *current* perceptions about the *future* and use them to move back into the *present* to estimate *current* values. And it is humans doing the perceiving.

## A MODEL FOR RISK

### Risk = The Uncertain Return on an Investment

If you buy a stock for $100, you can imagine its price going up to $110 for a return of 10%, or down to $90 for a return of −10%. The risk of the stock is reflected in the range of possible returns you can envisage.

### A Random Walk for Stock Prices

A company is a complex organism. How can one model the range of possible returns that a share of its stock might accrue? The Efficient Market Model's answer to this question is radical: ignore complexity! It hypothesizes that the market, anthropomorphically speaking, has used all available knowledge about the company to determine the stock price. Therefore the next change in the stock price will arise only from new information, which will arrive randomly and therefore be equally likely to be good or bad as far as the company's future returns are concerned.

Quantitatively, the EMM assumes that additional bits of information arrive uniformly, say every $\Delta t$ seconds.[8] Figure 5.1 illustrates how changes in the stock price occur in the model as a result of the arrival of news. By assumption, good news causes the company's value to increase by a fixed percentage; bad news causes the value to decrease. On average the stock price will grow, but with random up or down percentage fluctuations about the mean. The Efficient Market Model assumes that nothing wilder than these kinds of fixed fluctuations about the mean occurs (or is *imagined* to occur, because in finance, models are models of what is *expected*).

These apparently naïve up or down alternatives represent an attempt to model the essential risk of owning a stock. Repeated over

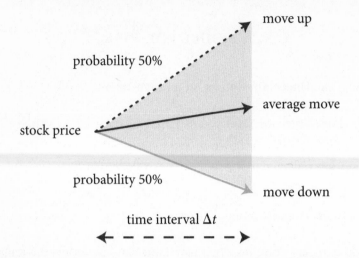

**Figure 5.1** A simple model for a stock's returns during an instant of time $\Delta t$. The stock has equal probabilities of random positive and negative returns about the average growth rate.

and over again for infinitesimally small time steps, they mimic the more or less continuous motion of stock prices in a plausible way, much as movies produce the illusion of real motion by changing images at the rate of 24 frames per second. Is the mimicry accurate as well as plausible? The answer to that question determines the difference between reality and fiction, between a theory and a model.

Figure 5.2 illustrates a stock price in the EMM. The price has undergone seven random up and down moves in (imagined) correspondence with the arrival of bits of good or bad news. Because three of the seven moves are down and four are up, the final stock price's up and down moves almost cancel, resulting in a net move up that is the result of only one bit of bad news, even though seven bits arrived. The net fluctuation about the mean return is much less than the seven-times-larger fluctuation one might have naïvely expected, because the ups and downs tend to cancel.

The pattern of small net fluctuations holds more generally. One

initial stock price

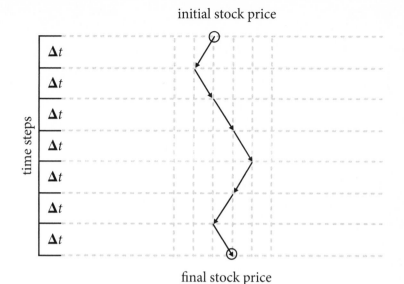

final stock price

**Figure 5.2.** An illustration of seven successive random moves in the stock price. Notice that after seven moves the stock price has moved only one net unit from its initial value.

can show that after $n$ random up or down moves the magnitude of the average return, whether net up or net down, is not $n$ times larger than one random move, but only $\sqrt{n}$ times as large.

Thus after 100 items of news will have arrived the stock price will most likely have drifted only about $\sqrt{100}$, that is, 10 uncertain price steps, up or down, away from its original value. After 1,000 time steps the average up or down movement in the stock price will have been only about 32 steps. And after one million time steps the average will be only 1,000 net up or down moves, pretty small compared with a million.

Because the number of bits of information $n$ that arrive during time $t$ is proportional to the time available, and because each bit moves the stock price about some average return (Figure 5.1), the stock price tends to grow as time passes. Its average return increases with time and can be written as $\mu t$, where the coefficient $\mu$ (the Greek letter mu, pronounced "mew") that multiplies the time elapsed is the average rate of growth. In contrast, the average fluctuation up or down *about* this growth rate can be written as $\sigma\sqrt{t}$, where the coefficient $\sigma$ (the Greek letter sigma) measures the size of the fluctuations, which, because of the tendency of the up and down moves to approximately cancel, increases only as the square root of time elapsed.

Figure 5.3 illustrates the assumed behavior of stock prices in the EMM for short intervals of duration $\Delta t$. This process is called *diffusion*. The stock price is expected to appreciate through time at some average rate $\mu$ per year, which finance aficionados call its *drift*. Part of the time it will grow more rapidly, part of the time less so. The intrinsic size of the yearly fluctuations in the drift is called the *volatility* $\sigma$ of the stock, and the net fluctuations increase slowly, proportional to the square root of time. More volatile stocks have larger $\sigma$'s and less

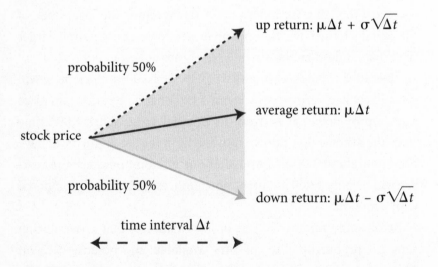

**Figure 5.3.** Stock price evolution during a short interval $\Delta t$ in the Efficient Market Model. The stock grows at an average rate $\mu$ with a volatility $\sigma$.

volatile stocks smaller ones, but whatever their volatility, the net uncertainty of their return increases with the square root of the time elapsed.

If you think of risk as the quantifiable uncertainty of imagined returns, then a key result of the EMM is that risk grows comparatively slowly, as the square root of time. Suppose you expect a stock to earn 30% in one year with a volatility of 10 percentage points. All circumstances being equal during the second year, you will correctly expect to earn double the return, 60% over two years. Counterintuitively, at least until you get used to the EMM, you should not expect the uncertainty of the returns to double to 20%. After two years the uncertainty in returns is only its square root, about 14 percentage points. Risk grows more slowly in the model, though not necessarily in life itself.

### Drift μ and Volatility σ: A Numerical Example

Suppose that $\mu = 5\%$ per year and $\sigma = 3\%$. This means that the stock price is expected to grow at an average rate of 5% per year with a volatility of 3 percentage points, meaning you can expect random up or down oscillations of ±3 percentage points per year about the drift of 5%. During some short periods, therefore, the stock price will grow at an annual rate of 8%, and for other short periods at a rate of 2%. The risk is that you will not know which rate is about to prevail.

A more volatile stock could have the same drift of 5% with a volatility of 10%, so that sometimes its price may grow at 15%, but at other times at −5%, its price decreasing rather than growing.

## RISK AND RETURN

The radically naïve assumption of the EMM is that the only type of risk stock prices suffer is the tame risk of diffusion, a risk that increases relatively slowly as the square root of time. The EMM allows no wilder price movements. In that case the behavior of stock prices is characterized entirely by the stock's expected return $\mu$ and its volatility $\sigma$.

The deepest question in finance is the relation between risk and return, or between $\sigma$ and $\mu$: if you expect a security to have volatility $\sigma$, how much average return $\mu$ should you expect? Let's begin by examining the risk and return of some common securities.

### Treasury Bills

The simplest, safest way to generate future cash from current cash is to buy a short-term, three-month Treasury bill. The purchase price you pay is loaned to the government, which in return promises to pay you a guaranteed rate of interest for three months and then return your principal. Because it is very unlikely that the U.S. Treasury will not be around to repay you three months later, the investment is close to riskless and therefore pays a low rate of interest. Its return serves as a benchmark for riskier securities, which must promise to pay more.

### Corporate Bonds

Lending money to a corporation (that is, buying a corporate bond) is riskier because corporations do sometimes go bankrupt. They must therefore promise to pay a greater rate of interest than the Treasury because their promise is more likely to be broken. And so, somewhat paradoxically, the more they must promise to pay, the more likely it is that they will fail to do so.

## Stocks

Suppose that, instead of lending money, you buy a share of company stock. Though investors have expectations of the magnitude of future returns and base their investment on it, the company makes no promise of how much the investment will actually yield. That will depend on fortune. If you sell the stock after three months, you will have to take what the market is willing to pay. In short, buying a stock has a highly uncertain outcome.

## Greater Return Means Greater *Average* Return

It's a platitude in finance that people demand greater return for exposing themselves to greater risk. But what does it mean to say you expect greater return when there is a chance you may not get it? It's a little subtle because you have to imagine the range of all possible returns, from good to bad. The expectation of greater return is an expectation of greater *average* return, averaged over all possible profits and losses as a result of your investment. Because a stock's return is prone to greater risk and consequent misfortune than that of a Treasury bill, investors expect a greater return, *on average,* from the riskier stock.

"On average" can be interpreted in (at least) two ways. The first kind of average is a *time average*: if you make a single investment and measure the returns it generates year by year over many years, the time average is the average of the successive yearly returns. The second kind of average is a *portfolio average*: if at one time you buy many distinct but roughly similar securities with the same risk, the portfolio average is the average of the returns over all securities with the same risk during the immediately following year.

Mostly in finance one assumes without proof that these two averages are equivalent—probably a false assumption. The financial world is not particularly stable, so long-term averages will depend on the prevailing environment, which, unlike that of the world of physics, isn't stable. Governments, administrators, and regulators change, booms and busts come and go, people modify their behavior based on recent failures and successes, and so the patterns of returns and their statistical distributions tend to change with time.

## THE ONE LAW OF FINANCE

### The Crucial Question: The Relation Between Risk and Return

The key question for any model or theory of finance is therefore: *What average return above the riskless rate should you expect to earn for accepting a given amount of risk?*

Notice that the question is about one's *expectations* of return rather than the returns that will actually be realized. Understand that expectations about the future are precisely a model. Notice also that realized and expected returns may not agree. When realized returns differ too wildly from expected returns, the world awakes, startled at the inadequacy of the models behind the prices. One of the lessons of the credit crisis that began in 2007 is that the models used to determine the value of mortgage securities did not include in their expectations the possibility of the catastrophic future scenarios that actually occurred.

### The Law of One Price: Similar Securities Have Similar Prices

To answer a general question requires having a principle. The only principle you can rely on in finance (and it's not always reliable) is the

Law of One Price: *If you want to know the value of one financial security, your best bet is to use the known price of another security that's as similar to it as possible.*

When we compare it with almost everything else in economics, the wonderful thing about this law of valuation by analogy is that it dispenses with utility functions, the undiscoverable hidden variables whose ghostly presence permeates economic theory. Financial economists like to recast the Law of One Price as the more pedantically named Principle of No Riskless Arbitrage: *Any two securities with identical future payoffs, no matter how the future turns out, should have identical current prices.* This Law of One Price embodies the common sense that the author of "the fundamental theorem of finance" was trying so hard to convey but expressed so unclearly.

In the imaginary world of the Efficient Market Model, a stock's price movements are completely characterized by its expected return $\mu$ and its volatility $\sigma$. All that differentiates one stock from another is their respective values of $\mu$ and $\sigma$—nothing else. Two securities with the same $\mu$ and $\sigma$ are therefore similar. The Law of One Price combined with the EMM then leads to the following conclusion: *Any two securities with the same foreseeable volatility should have the same expected return.*[9]

There is a shallow resemblance between the Law of One Price in finance and Einstein's principle of relativity in physics, presented in the table below.

| Principle of Relativity | Law of One Price |
|---|---|
| All observers, irrespective of their speeds, should, from their observations, deduce the same equations for the laws of nature. | All securities with identical risk, irrespective of their nature, should provide the same expected return. |
| Light is special; nothing can move faster than light. | Riskless bonds are special; nothing can have less risk than a riskless bond. |

The Law of One Price is not, however, like the principle of relativity. It is not a consistent law of nature. It is a general reflection on the practices of fickle human beings, who, when they have enough time, resources, and information, would rather buy the cheaper of two similar securities and sell the richer, thereby bringing their prices into equilibrium. The law usually holds in the long run, in well-oiled markets with enough savvy participants. In crises, however, duress forces people to behave in what looks like irrational ways, and even in normal times there are persistent shorter- or longer-term exceptions to the law.

How do you use the Law of One Price to determine value? If you want to estimate the unknown value of a "target" security, you must find a "replicating" portfolio, a collection of liquid securities that, collectively, are similar to it, that is, have the same future payoffs as the target, *no matter how the future turns out*. The target's value is then simply the price of the replicating portfolio, which, being liquid, will have easily available prices.

This is where models enter the picture. It takes a model to demonstrate similarity, to show that the target and the replicating portfolio have identical future payoffs *under all circumstances*. To demonstrate the identity of future payoffs, you must (1) specify what you mean by "all circumstances" for each security, and (2) find a recipe for creating a replicating portfolio that, in each future scenario, will have payoffs identical to those of the target.

### Using the Law of One Price to Value Apartments

Here's a simple financial model that illustrates most of the characteristics of more sophisticated models. Consider the problem of estimating the value of a grand eight-room penthouse apartment on Park Avenue in New York City. The most direct way is to put the apartment on the market and look at the offers. Next best is finding a recently sold, similar apartment in the same neighborhood and using its sale price as the approximate value of this apartment.

Suppose that there have been no recent sales of similar apartments, and that you can obtain only the recent sale price of one of many vanilla two-room apartments in Battery Park City. These apartments are liquid and change hands frequently. You will have to value the Park Avenue apartment relative to the price of the Battery Park City unit. You can approximately *replicate* the penthouse, in your imagination, by combining several Battery Park City units into one larger unit the size of the penthouse. Suppose that requires about seven Battery Park City apartments. Then your first estimate is that the Park Avenue penthouse should be worth seven times the Battery Park City apartment.

This model assumes that the price per square foot of apartments is constant, irrespective of apartment details. It's a *model* because it assumes that square footage is all that matters in estimating the value of the apartment. That's clearly not true. First, the quality of construction matters. Second, in New York City very large apartments are scarce and sell at a premium. Third, a Park Avenue location is more exclusive and hence more desirable.[10] Fourth, there are more staff and doormen per capita in a Park Avenue cooperative building. Et cetera.

In this model price per square foot is what quants call an *implied parameter*. It's "implied" in the sense that it's not the literal construction cost per square foot, but rather the *implied* price per square foot, obtained by quoting the market price of the entire apartment and all its accoutrements and bundled services in terms of the price per square foot.

The price of the replicating portfolio is therefore an underestimate. With some practical experience you can make rule-of-thumb corrections for size, location, park views, natural light, and apartment detailing to adjust the initial estimate of a factor of seven. This is the way most practical financial models work. You use similarity to get off the ground, comparing the characteristics of something you don't know with those of something you do. In the apartment valuation model, square footage captures a major part of the variation in apartment prices. Then you make commonsense heuristic adjustments for other important factors that lie outside the model. Models for pricing stocks, bonds, and options work the same way.

## THE CONCLUSION: EXCESS RETURN IS PROPORTIONAL TO RISK

The Law of One Price and the EMM both require that securities with the same risk must have the same return. But *how much* return? Because there is only one variety of risk in the EMM, the volatility $\sigma$ of the stock price's diffusion, you can easily engineer any amount of risk from one sample of it. Here's the logic.

Suppose you begin with a stock that carries a particular volatility $\sigma$ and from which you expect a return $\mu$. Then you can easily create a portfolio with less risk by diluting the stock with a riskless bond. But because the return of a portfolio is the weighted average of the returns of its ingredients, you know what return to expect for it. Hence, by the Law of One Price, you know what return to expect for all securities with the same risk as that of the portfolio. In short, if you know the risk and return for one stock, you know the risk and return for all. Here is the precise recipe for diluting the mixture:

*Ingredients:* Prepare one dollar's worth of a stock with volatility $\sigma$ and expected return $\mu$, while keeping ready for use one dollar's worth of a riskless bond with zero volatility and guaranteed return $r$.

*Preparation:* Mix together thoroughly a fraction $w$ of the stock with a fraction $1-w$ of the riskless bond.

The mixture is worth exactly $1 too, but it is riskier than the bond and less risky than the stock.

By doing a little algebra you can see that the volatility of the mixture is

$$w\sigma$$

because only the stock part of the mixture carries risk. By equally straightforward algebra the expected return of the mixture is the weighted average of the return of the riskless bond and the stock:

$$w \times \mu + (1 - w) \times r = r + w(\mu - r)$$

Focus on the terms involving $w$ in these expressions and you will notice that adding an *extra $w\sigma$* amount of risk *to a riskless bond* wins you an *extra $w(\mu - r)$ of expected return*, over and above the riskless bond's return $r$. Thus by the Law of One Price we obtain the Law of Proportionality of Risk and Return:

If you expect return $\mu$ for risk $\sigma$, then you should expect an excess return $w(\mu - r)$ from any security with risk $w\sigma$.

By choosing different fractions $w$ of the stock in the mixture, you can create risk and expected return of any size. The more $w$, the more risk, the more expected return.

**A Numerical Illustration**

Suppose the benchmark return of a riskless bond is currently 2% per year. Suppose the market expects a return of 7% on a stock with a volatility of 20%. From this you can figure out the return for any volatility, as follows: The stock with a volatility of 20% provides an excess return of 7% − 2% = 5%. Therefore a stock with half its volatility, that is, 10%, must provide half as much expected return, that is, 2.5% Adding this to the riskless bond's return of 2%, you should expect a return of 4.5% on a stock with volatility of 10%—if the EMM is correct.

**The Sharpe Ratio**

According to the Law of Proportionality of Risk and Return, a risk $w\sigma$ must provide an excess return $w\,(\mu - r)$. Put differently, this means that for any security the ratio of excess return to risk is always the same, since each is proportional to the weight $w$.

Finance theorists like to write the Law of Proportionality of Risk and Return entirely in Greek symbols to make it seem oracular:

$$\frac{\text{excess return}}{\text{risk}} \equiv \frac{\mu - r}{\sigma} = \lambda \qquad \text{Sharpe ratio}$$

This states that the ratio of excess return to risk is a "universal" constant, denoted by the Greek letter $\lambda$ (lambda), which is referred to as the Sharpe ratio, after William Sharpe, who first began to make use of this concept in the mid-1960s. It measures the bang you hope to get for your buck of risk, and is therefore also called the *risk premium*.

In the framework of the EMM it follows that, on average, by taking greater risk you should automatically win greater reward. You don't have to be smart to make more; you just need to invest in a security with higher volatility. But you will earn a higher reward only *on average*, not every time. That's why it's risky. So how do you tell if you are a smart investor? According to the EMM the test is the value of your portfolio's Sharpe ratio. If your portfolio contains securities whose realized Sharpe ratios turn out to be greater than the market's average Sharpe ratio, you have been smart. Of course, only if the EMM weren't quite right could you hope to find exceptional Sharpe ratios.

The theory of efficient markets has become so much a part of accepted market lore that professional money managers measure and report their Sharpe ratios with the hope of demonstrating their talent.

---

### Judging Investments by Their Sharpe Ratios

Suppose that over the past 10 years a security A has produced an average annual return $\mu_A = 15\%$, with high returns of 20% and low returns of 10%, so that the annual volatility $\sigma_A$ is 5%, corresponding to return fluctuations of ±5 percentage points about the mean.

Suppose also that the riskless rate $r$ for one-year Treasury bills during this time was 5%. Then the excess return for security A was $\mu_A - r = 15\% - 5\% = 10\%$ annually.

The annual Sharpe ratio was $\dfrac{\mu_A - r}{\sigma_A} = \dfrac{10}{5} = 2$. This means you have earned two units of excess return for every one unit of risk each year.

---

Now suppose you are confronted with a security B that earned an average return equal to 20% per year over the same period, with lows of 5% and highs of 35%. How does that compare with having invested in A?

The excess return for B over and above the riskless rate was $\mu_B - r = 20\% - 5\% = 15\%$, which *looks* better than the excess return of 10% for A. But the volatility $\sigma_B$ of B, corresponding to half the range between its high and low returns, was $\frac{1}{2}(35\% - 5\%) = 15\%$. The Sharpe ratio for B was therefore $\frac{\mu_B - r}{\sigma_B} = \frac{15}{15} = 1$.

B looks better in terms of returns, but A is better in terms of return per unit of risk taken, which, *if there is only one kind of risk,* is what matters. If there is only one kind of risk, you are better off exposing yourself to it through security A than through security B.

In this example we have calculated historically realized Sharpe ratios, looking backward rather than forward in time.

## AN ASIDE: THE PLEASURE PREMIUM

The Sharpe ratio $\lambda$ tells you how much reward you should expect from a given chunk of risk. Risk is a fundamental quality, something you can choose to expose yourself to or try to avoid, a primitive that cannot be reduced to something simpler. The demand for risk reflects psychology and the physiology beneath it. Humans will expect different degrees of return from a given amount of risk at different times. When they become more risk-averse and demand more return from accepting the same risk, stock prices fall and the Sharpe ratio rises; conversely, when they become willing to accept less return, the ratio will fall and prices will rise.

The desire for risk is a biological primitive, comparable to the desire for deliciousness in food. The deliciousness of food is an independent quality, distinct from its caloric and nutritional value. Chefs in restaurants can charge more for greater expected *pleasure* and deliciousness. If you wanted to be foolishly quantitative, you could define the *pleasure premium* as the amount of dollars diners are willing to pay, *over and above the bare cost of the ingredients themselves,* for deliciousness. You can make the following analogy:

| Risk of a Security | Deliciousness of a Meal |
|---|---|
| Expected return on security: $\mu$ | Cost of meal: $C$ |
| Return of riskless bond: $r$ | Cost of ingredients: $I$ |
| Excess return: $\mu - r$ | Excess cost: $C - I$ |
| Risk: $\sigma$ | Deliciousness:[11] $\Omega$ |
| Risk premium $\dfrac{\mu - r}{\sigma}$ is the excess return per unit of risk. | Pleasure premium $\dfrac{C - I}{\Omega}$ is the excess cost per unit of deliciousness. |

The pleasure premium will vary from day to day; in flush times diners may be willing to pay more for expected deliciousness; in bad times they may avoid deliciousness and stick to nutrition. In a quantitative model the pleasure premium $\Omega$ could play a large part in determining luxury food prices.

In theory securities sell on the basis of expected return, and in theory delicacies could sell on the basis of expected deliciousness. But there is a difference between preprandial expected deliciousness and postprandial realized deliciousness, and if the mismatch is too great, the chef may be patronized no more. Similarly, the expected return of a security differs from its realized return, and if the mismatch is too great, the security will be shunned, as has happened with collateralized default obligations. But whereas expected return and realized return are completely distinguishable, expected pleasure is often a pleasure itself.

What is the appropriate value for $\Omega$, the excess cost per unit of expected deliciousness? No theory can dictate what a sybaritic society should pay for pleasure. One might theorize that people pay the same amount per unit of deliciousness no matter what food delivers it. Similarly, no financial theory can dictate what return an investor should expect in exchange for taking on risk; that too depends on appetite and varies with time.

I am of course simplifying things by treating all pleasure as equivalent; there is more than one kind of deliciousness. There are sweetness and tartness, spiciness and blandness, smoothness and lumpiness, all of them different types of gustatory pleasure. And I have ignored other kinds of mentionable and unmentionable bodily pleasures, which have their pleasure premiums too. Similarly, there is more than one kind of risk and more than one kind of risk premium: stock risk and bond risk and currency risk and commodity risk and slope-of-the-yield-curve risk; and within the universe of stocks there is sector risk—health risk, technology risk, consumer durables risk, et cetera.

In physics the values of the fundamental constants (the gravitational constant $G$, the electric charge $e$, Planck's constant $h$, the speed of light $c$) are apparently timeless and universal. I doubt there will ever be a universal value for the risk premium.

## THE EMM AND THE BLACK-SCHOLES MODEL

The best model in all of economics is the Black-Scholes Model for valuing options on stocks, an ingeniously clever extension of the EMM published in 1973 by Fischer Black and Myron Scholes.* I spent my first two years at Goldman Sachs, 1986–1987, working with Fischer Black on an extension of this model to valuing options on bonds,[12] and I devoted 1993–1994 to working on an extension of Black-Scholes to stocks with variable volatility.

Stocks are commonplace, but a call option on a stock is more arcane, and prior to Black and Scholes no one had a plausible clue for how to value it. A call option works as follows. If you buy a one-year call on Apple, you have the right to buy one share of Apple one year later at an agreed-upon price, say $350. The value of the option one year hence will depend on the value of Apple stock on that day. For example, if the share price turns out to be $370, the option will be worth exactly $20; if the share price ends up being $350 or less, the option will be worth zero. The call option is a bet that the stock price will rise.

Assuming that Apple stock moves randomly in accord with the precepts of the EMM, then the more volatile the stock, the more likely its price will be to wind up above $350, and therefore the more the bet is worth. Because the option is a risky bet on the risky stock, it's riskier than the stock, and yet its risk depends on the risk of the stock itself.

Black, who began by studying physics, was a firm believer in arguments based on equilibrium. In physics, if you mix together two gases at different temperatures, they will settle into equilibrium when

*Myron Scholes together with Robert C. Merton, who derived a different proof of the Black-Scholes formula and developed much of the elegant mathematics associated with options pricing, received the 1997 Nobel Memorial Prize in Economics for their work on the model. Fischer Black died two years earlier.

the heat flowing out of the first gas and into the second is the same as the heat flowing out of the second and into the first. Black applied analogous logic to options and stocks. If you consider a market with two securities, Apple stock and Apple call options, then you have two ways to expose yourself to Apple risk: via the stock or via the option. The market will be in equilibrium when the price of the option and the price of the stock are such that each security's Sharpe ratio—each security's expected excess return per unit of its risk—is the same for both, so that investors will have no reason to prefer one security over the other as a route to taking on Apple risk. Until that is the case, investors will preferentially buy the more efficient security and sell the less efficient one until their prices adjust to demand and they eventually provide the same risk premium. That's equilibrium.

By equating the Sharpe ratio of the stock and the Sharpe ratio of the option, Black and Scholes were able to derive and, a few years later, solve an equation for the model value of the call option. Furthermore, because the option must at all times have the same risk premium as the stock, you can replace the option at any instant by an equivalent investment in stock. The Black-Scholes Model tells you exactly how much stock you need to replicate the option's risk at any instant, and thus, if you know the stock price, what the option is worth at any instant. It's like a recipe that tells you how to make fruit salad (an option) out of fruit (stocks and bonds) and hence, by the Law of One Price, what the fruit salad is worth.

Before Black and Scholes and Merton no one had even guessed that you could manufacture an option out of simpler ingredients. Anyone's guess for its value was as good as anyone else's; it was strictly personal. The Black-Scholes Model, even more than the EMM that engendered it, revolutionized modern finance. Using Black and Scholes's insight, trading houses and dealers could value and sell options on all sorts of securities, from stocks to bonds to currencies, by synthesizing the option out of the underlying security.

The Black-Scholes Model is only a model, and so the synthesis is of course imperfect in practice. Nevertheless, Black-Scholes works for options much better than the EMM works for stocks. The EMM is at best a simplistic model for risk that neglects the subtleties of stock price behavior. But Black's great idea, that an option and a stock should share the same risk premium when markets are in equilibrium, is close to a more general truth. Irrespective of subtleties, the risk of the stock and the risk of the option are sufficiently related so that equating their risk premiums gives a sensible constraint on their relative prices. That makes the Black-Scholes Model robust; it is a rarity among financial models.

In contrast, the EMM works much less well for stock valuation, because stock prices suffer risks more diverse and wild than those associated with diffusion. There are many other huge risks—among them, the enthusiasm of crowds that can make entire stock markets rise, the fear that can make them crash, liquidity that can dry up, counterparties that can all fail together in a crisis—that the EMM ignores and that can therefore invalidate many of its results. This is what happened in the great financial crisis.

## THE CAPITAL ASSET PRICING MODEL

The Capital Asset Pricing Model, developed by Jack Treynor, William Sharpe, Jan Mossin, and John Lintner in the early 1960s, is an extension of the EMM that more realistically takes account of the risk not just of single stocks but of the entire stock market. Finance aficionados refer to the model affectionately as CAPM ("Cap Em"). Though economists regard CAPM as the triumph of so-called modern portfolio theory, and though CAPM engendered the Black-Scholes Model, it isn't nearly as robust, and therefore isn't nearly as useful.

The EMM as described thus far focuses on individual stocks and assumes that their prices do nothing more dramatic than diffusion.

CAPM extends the EMM by recognizing that stocks are part of a larger market. In practice, stocks tend to move in tandem, most of them going up when the market goes up and going down when the market goes down. One reason for this co-movement is Keynes's now clichéd but nevertheless accurate observation that it requires "animal spirits" to take risk: investors pile into stocks when the world looks good, and rush to safety when things go bleak. To make the model more realistic, the founders of CAPM added this herding tendency to the EMM.

CAPM recognizes two kinds of risk embedded in each stock: the overall *market risk* that affects everything, and the stock's *idiosyncratic risk* independent of the market. Market risk affects every stock and is therefore unavoidable once you decide to buy any stock at all, whereas idiosyncratic risk can be different for each stock, its own individual personality trait. When you buy even a single stock you face both its market risk and its idiosyncratic risk. The insight of CAPM is that because market risk is unavoidable, it is the only genuine risk you have to take if you want to play the stock market, and therefore the only one you should expect to be rewarded for. The capitalist mantra "More risk, more return" according to CAPM applies to market risk only.

Idiosyncratic risk is avoidable: it can be diminished by *diversification,* the assembly of a portfolio of many stocks whose idiosyncratic risks are unrelated and therefore tend to cancel each other out. (This tendency to cancel is similar to the tendency of independent random up and down stock price moves in Figure 5.2 to cancel, so that the idiosyncratic risk of a portfolio of $n$ stocks grows only as fast as $\sqrt{n}$ .) All that afflicts a stock in a (theoretically) diversified portfolio of many stocks is therefore the market risk it carries.

CAPM characterizes each stock's unavoidable market risk by a statistic, $\beta$ (beta), the ratio of its individual market risk to the risk of the entire market. The $\beta$ of each stock describes its tendency to herd with the crowd, and different stocks have different measurable betas.[13] The greater the beta of a stock, the more it responds to a market move.

Exploiting the idea that any stock's excess return is due only to the market risk it bears leads to the following conclusion for markets in equilibrium: *All of a stock's excess expected return is equal to its beta times the market's excess expected return.*[14] According to CAPM, the more the stock's price moves with the market, the more you should expect to earn.[15]

Utility stocks, for example, are stodgy and safe, less risky than the market, because utilities supply gas and electricity that are necessities for survival: animal spirits or no animal spirits, everyone is going to buy heating. Consolidated Edison (stock ticker ED), whose fondly remembered 1960s motto "Dig We Must!" advertised its frequent destruction of New York City roads as a utilitarian necessity for the benefit of New Yorkers, has a low beta of about 0.3. So (if you believe CAPM) you should expect only about one third of the market's excess return from Con Ed.

In contrast, Apple (stock ticker AAPL), a company that lives by inventing and creating demand for usefully vogue products you can survive without, depends on animal spirits. The beta of Apple is about 1, much higher than that of Consolidated Edison and about the same as that of the market in general, and so, according to CAPM, you should hope for three times as much excess return from Apple as from Con Ed, about as much as you get from the market. The following box shows that this hasn't been the case.

---

### A Simple Test: Apple and the S&P 500

Here is a plot of the prices of Apple (AAPL) and the S&P 500 (ticker ^GSPC) between September 16, 2009, and September 13, 2010. (All data and figures are taken from www.wolframalpha.com.)

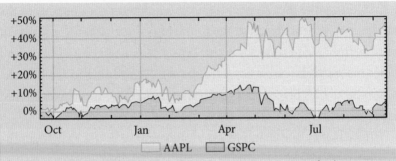

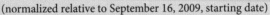

(normalized relative to September 16, 2009, starting date)

Here is a scatter plot of the daily returns of Apple and the S&P 500 over the same period.

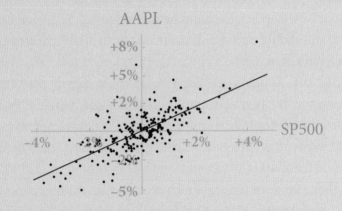

Let's see how well the CAPM worked for Apple stock during that year.

During the one-year period, Apple's risk relative to the market was $\beta = 1.05$.

Let's be generous and say that the annual riskless rate was about 1% at the start of that year, so that $r = 1\%$ in the equation in note 14.

CAPM states that, written using algebra,

$$(\mu - r) = \beta\,(\mu_M - r).$$

Strictly speaking, CAPM claims that this must hold for *expected* returns. But no one accurately knows what people expected a year ago, and so it is common to check the validity of CAPM for *realized* returns. During that period, AAPL returned about 55%, while the S&P 500, a good proxy for the market, returned about 11%.

The value of the left-hand side of the equation is

$$\mu - r = 55 - 1 = 54\%.$$

The value of the right-hand side of the equation is

$$\beta\,(\mu_M - r) = 1.05\,(11 - 1) = 10.5\%.$$

The value of the left-hand side is about 44 percentage points greater per year than the value of the right-hand side. CAPM didn't work well at all during this period.

The predictions of CAPM in this example were violated by Apple to the tune of about 44 percentage points of excess return over and above the value required by CAPM. Aficionados use the Greek letter $\alpha$ (alpha) to refer to the amount by which a stock's actual return exceeds the fair beta-inspired CAPM value.

Why isn't alpha zero? A kind way to explain the mismatch is to say that CAPM holds over the long run, on average, with fluctuations from year to year and from security to security. After all, CAPM is meant to describe the market in equilibrium, and there are periods when the market must be between equilibria. To test it you should test it statistically, not for one stock at one time. The particular period considered in the example should be regarded as a statistical fluctuation.

A still kinder way to explain it is to say that CAPM is incomplete and needs extending. Market risk is *not* the only herding risk. Savvy investors are always trying to make sense of the market's behavior, and rightfully perceive stocks as belonging to groups smaller than the entire market, groups whose constituent stocks tend to be bought or sold as a class.[16] Large capitalization stocks, "small caps," value stocks, growth stocks—each is an ensemble of companies whose fortunes tend to move together. Looking more closely, we see that computer stocks move together too, because certain kinds of news are relevant to the computer industry but relatively insignificant for, say, health care stocks. Each of these additional groupings and the factors that represent them are additional types of risk that can lead to more complex versions of CAPM.

An unkind way to look at CAPM is to say that it's not very good. Newton's law it ain't.

## CAPM's Infiltration

Despite its lack of great success, CAPM has infiltrated the language of finance, even nonacademic finance. I didn't fully understand how deeply until I recently tried to use Bloomberg, Yahoo, and Google to find the volatilities of Apple and the S&P 500 to create the box above. To my astonishment there was no easy and direct way to obtain a stock's volatility $\sigma$. All those websites give you easy access only to the stock's beta, the amount of market risk carried by the stock that is the hallmark of CAPM's view of the world. It is a sign of the political power of models that commercial websites publish the value of beta, a parameter in a model that doesn't work that well, but not the more fundamental and model-neutral volatility statistic $\sigma$.

But CAPM has had a beneficial impact, having being responsible for introducing into finance as metaphors the notion of alpha and beta. Alpha and beta represent the sources of return. Beta refers to the return earned for simply entering the market dumbly. Getting beta is easy: all you have to do is buy a diversified portfolio of stocks without worrying about their individual attributes. Alpha, in contrast, represents skill, the return generated by being smarter than the hordes, by picking better-than-average stocks, or picking ordinary stocks at the right time, as with Apple in the example above. Inspired by CAPM, investors now ask themselves whether their manager is providing merely dumb beta or smart alpha. Alpha is worth paying a fee for, and it's supposed to be what hedge funds provide. Beta should be cheap: anyone can whistle.

## EMM: A THEORY OR A MODEL?

I showed that expected return is proportional to risk by using just two principles: (1) you should expect equal returns from equal risks, and (2) a stock's risk is solely the volatility of the diffusion illustrated in Figure 5.3. The first principle is pure theory and hard to argue with: If two securities truly have the same risk, how could you not expect the same return from them? But that's an expectation. In life, expectations aren't necessarily fulfilled. The second assumption is pure model. The EMM's picture of price movements goes by several names: a random walk, diffusion, and Brownian motion. One of its origins is in the description of the drift of pollen particles through a liquid as they collide with its molecules. Einstein used the diffusion model to successfully predict the square root of the average distance the pollen particles move through the liquid as a function of temperature and time, thus lending credence to the existence of hypothetical molecules and atoms too small to be seen.

For particles of pollen, the model is also a theory, and pretty close to a true one. For stock prices, however, it's *only a model*. It's how we

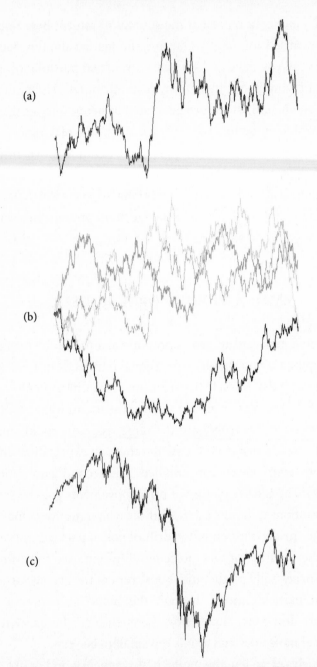

**Figure 5.4.** (a) A single stock path simulated via a random walk. (b) Four typical simulated stock paths. (c) An actual four-year path for the level of the S&P 500 index.

*choose to imagine* the way changes in stock prices occur, not what actually happens. It is naïve to imagine that the risk of every stock in the market can be condensed into just one quantity, its volatility σ. Risk has too many aspects to be accurately captured by that one number.

### The EMM Isn't Wild Enough

Figure 5.4 compares the paths of stock prices generated from the EMM model with the level of the S&P 500 from late 2006 to late 2010. The apparently naïve either-up-or-down model does superficially mimic the riskiness of a stock's price.

But only more or less. The mimicry fails because the stock paths in the model are just too smooth when compared with the observed movements of actual stock prices. When examined closely, the stock price trajectory in Figure 5.4c is jerkier than those in Figure 5.4b. Actual stock prices are more wildly random than those of the model, as becomes very obvious during stock market crashes, when stock prices cascade downward in giant leaps, and volatility spikes beyond belief.

It's not surprising, then, that CAPM doesn't correctly account for the returns on investments. CAPM may hold better in undramatic, liquid markets where informed investors determine prices. But the model's assumptions fail badly during times of panic, fear, and limited liquidity. CAPM is a useful way of thinking about a model world that is, quite often, *far from* the world we live in.

### Conceptual Mismatches

There are also glaring conceptual flaws. In contrast to the model's assumptions, the news that drives stock prices doesn't always arrive in small, steady increments. Sometimes something very important happens suddenly and discontinuously. The demand for a stock can become so great that its price leaps up; more commonly the market's panic to sell can be so intense and contagious that the prices of all stocks crash downward, reflecting fear rather than the ideal of rational information and response. These kinds of events are not rare, yet they don't fit into the framework of the smooth random walk economists like to focus on. And it is these events that provide much of the true risk as well as the reward of investing. The Efficient Market Model's price movements are too constrained and elegant to reflect the market accurately.

## THE UNBEARABLE FUTILITY OF MODELING

To use the Law of One Price that underpins financial modeling, one must show that a target security and its replicating portfolio have identical future payoffs under all circumstances. Most of the mathematical complexity of modeling in finance involves the description of the range of future behavior that composes all circumstances. Trying to specify this always reminds me of the 1967 movie *Bedazzled,* starring Peter Cook and Dudley Moore. In this comic retelling of the legend of Faust, Moore plays a short-order cook at a Wimpy's chain restaurant in London who sells his soul to the devil in exchange for seven chances to specify the circumstances under which he can achieve his romantic aims with the Wimpy's waitress he desires. Each time the devil asks him to specify the romantic scenario in which he believes he will succeed, he fails to specify it sufficiently precisely. He says he wants to be alone with the waitress in a beautiful place where they are both in love with each other. He gets what he wants: with a snap of the devil's fingers, he and his beloved are instantly transported to a country estate where they are wild about each other. But he is a guest of the owner of the estate, her husband, whom her high principles will not allow her to betray. Finally, when he has only one chance left, he wishes for them to be alone together and in love with each other in a quiet place where no one will bother them. He gets his wish: the devil makes them both nuns in a convent where everyone has taken a vow of silence.

The difficulty of the hopeful would-be lover is the same difficulty we face when specifying future scenarios in financial models; as does the devil, markets eventually outwit us. The devil is indeed in the details. Even if markets are not strictly random, their vagaries are too rich to capture in a few short sentences or equations.

Not all financial modelers are stupid or ignorant of the follies of using the EMM. One can easily invent more complicated models of risky stock prices that incorporate violent moves and ferocious out-bursts of risk. But in using such models one gives up simplicity for a still imperfect but more complex model that doesn't necessarily do better. As with earthquakes, it may be wiser to remember that catas-trophes much more dramatic than those the random-walk model envisages occur quite often. It is better to ensure that one owns a portfolio that will not suffer too badly under disastrous scenarios than it is to try to estimate the probability of destruction.

So die the dreams of financial theories. Only imperfect models remain. The movements of stock prices are more like the movements of humans than of molecules. It is irresponsible to pretend other-wise.

CHAPTER 6

# BREAKING THE CYCLE

*Caught in a fiendish cage* • *Avoiding pragmamorphism* • *The great financial crisis and the abandonment of principle* • *The point of financial models* • *Be sophisticatedly vulgar* • *Let the dirt be visible* • *Beware of idolatry* • *The modelers' Hippocratic oath* • *We need free markets, but we need them to be principled* • *Once in a blue moon, people stop behaving mechanically*

"Alas," said the mouse, "the world is growing smaller every day. At the beginning it was so big that I was afraid, I kept running and running, and I was glad when at last I saw walls far away to the right and left, but these long walls have narrowed so quickly that I am in the last chamber already, and there in the corner stands the trap that I must run into." "You only need to change your direction," said the cat, and ate it up.

—Franz Kafka, "A Little Fable"

## THE PERFECT CAGE

When I opened the refrigerator door one morning, I found a moth flying furiously around the interior. It had somehow managed to get inside, and was now fatally attracted to the refrigerator light. I tried to chase it out by flapping my hand behind it, but to no avail. Every

189

time I opened the door the light came on and the moth circled it frantically, unwilling to leave. Every time I closed the door it got dark and cold inside, not at all what the moth had stayed for, but because the door was closed it couldn't get out.

Finally I found a solution. I remembered the spring-loaded button in the hinge of the refrigerator door that controls the bulb inside. I opened the door and held down the button with my finger. The inside went dark and the moth headed out for the more stable light of the kitchen.

It was a fiendish cage, perfect for moths, whose tropism made them free to leave whenever the door was opened, but who wanted to leave only when it was closed.

> It was a miracle of rare device,
> A sunny pleasure-dome with caves of ice!
> —Samuel Taylor Coleridge, "Kubla Khan"

## THE MYSTERIES OF THE WORLD

The world of Substance is full of puzzles. Theories offer us the most successful and accurate way to describe the physical world. They are deep and difficult to discover; they require verification but no explanation; they are right when they are right. Models, however, live in the shallows and are easier to find; they require explanation as well as verification. We need both types of understanding.

The most successful theories humans have created are concerned with Extension. There are few genuine theories in the realm of Thought, and none that is quantitatively accurate. There we rely mostly on metaphors and models. Financial value, situated more centrally within Thought than Extension, is therefore less inclined to yield to mathematics or science; there are no isolated social systems on which to carry out the repeated experiments the scientific method requires, and so it is hard to study the regularities that might reveal the putative laws that govern them.

Given the success of mathematics in dealing with Extension, it has become tempting to treat Thought as though it were a kind of Extension too. Most models in the social sciences give in to what I like to call *pragmamorphism*,[1] by which I mean the naïve tendency to attribute the properties of things to human beings. Among the exceptions are Spinoza's theory of the emotions and Freud's theory of the psyche. Pragmamorphism characterizes the approach that finance has taken for the past 50 years.

It's pragmamorphic to equate psychological states with their material correlates, to equate PET scans with emotion. It's similarly pragmamorphic to assume the existence of a utility function in economics. It is clear that people have preferences. But is it clear that there is a function that describes their preferences? Pragmamorphic models are a valiant try, perhaps the best we can do quantitatively. But we should remember that we are being pragmamorphic when we use them, and that humans will necessarily lie outside the models' boundaries.

## MODELS THAT FAILED

People expect either too much or too little from financial models. To use them appropriately requires some degree of vacillation. One must begin boldly but expect little. If a little success is actually obtained, one must grow greedy. Then, when one has gone only a little too far, desist.

I was motivated to write this book by the global financial crisis that began in 2007. After more than 20 years of hubris, models collapsed. At the end of the cold war we imagined a future with no more history, a smooth stroll into the sunrise accompanied by democracy, privatization, and free markets. It hasn't worked out that way. Authoritarian versions of capitalism have spread. Privatization has become oligarchy. The gaps between rich and poor, managers and workers, and owners and employees have widened. Economic models have misfired and financial models have proved to be enormously inaccurate. More recently the prescribed cure of a Keynesian stimulus to jump-start spending and employment has had only a muted effect. Low interest rates, the Federal Reserve's cure for past crises and the progenitor of future ones, are being prescribed again. Lessons have not been learned.

I wasn't surprised by the failure of economic models to make accurate forecasts. Any assurance economists pretend to with regard to cause and effect is merely a pose or an illusion. They whistle in the dark while they write their regressions that ignore the humans behind the equations.[2] I was similarly unsurprised by the failure of financial models. Financial models don't forecast; they transform one's forecasts of the future into present value. Everyone should understand the difference between a model and reality and be unastonished at the inability of one- or two-inch equations to represent the convolutions of people and markets.

What did shock and disturb me was the abandonment of the principle that everyone had paid lip service to: the link between democracy and capitalism. We were told not to expect reward without risk, gain without the possibility of loss. Now we have been forced to accept crony capitalism, private profits and socialized losses, and corporate welfare. We have seen corporations treated with the kindness owed to individuals and individuals treated pragmamorphically, as things.

When models in physics fail, they fail precisely, and often expose a paradox that opens a door. When models in the social sciences fail, they fail bluntly, with no hint as to what went wrong and no clue as to what to do next. With no way forward, people try to restore the status quo ante at any cost.

## WHAT IS TO BE DONE?

Financial models aren't going to disappear. Data alone have no voice. Theorizing and modeling are what humans do and will continue to do. So how do we use models wisely and well?

I have explained that there are no genuine theories in finance. In physics, Newton's laws and Maxwell's equations are facts of nature, but the Efficient Market Model's assumption that stock prices behave like smoke diffusing through a room is not even remotely a fact. Financial models are always metaphors.

### WHAT IS THE POINT OF A MODEL IN FINANCE?

It takes only a little experience to see that the point of a model in finance is not the same as the point of a model in physics. In physics one wants to predict or control the future. In finance one wants to determine present value and goes about it by forming opinions about the future, about the interest rates or defaults or volatilities or housing prices that will come to pass. One uses a model to turn those opinions about the future into an estimate of the appropriate price to pay today for a security that will be exposed to that imagined future. I illustrated this in my model for valuing a penthouse in chapter 5, where, by assuming that price per square foot was the same for all apartments, I estimated the price of an eight-room Park Avenue penthouse by replicating it out of two-room apartments in Battery Park City. This process is characteristic of more general financial models.

Models are useful in finance and here are some of their major benefits.

**Models Facilitate Interpolation**

One uses financial models to interpolate or extrapolate from the current known prices of liquid securities to the estimated values of illiquid securities—relating the unknown value of the Park Avenue apartment to the known prices of those in Battery Park City, for instance. Similarly, the Black-Scholes Model proceeds from a known stock price and a riskless bond price to the unknown value of an option, which is a hybrid of stock and bond.

No model can be correct—a model is not a theory—and no replication is truly accurate. But models can be immensely helpful in calculating initial estimates of value.

**Models Transform Intuition into a Dollar Value**

The apartment-value model transforms intuitive knowledge about price per square foot, now or in the future, into the value of the apartment. It's easier to begin with an estimate of price per square foot because that quantity captures so much of the variability of apartment prices. Similarly, it's far easier to convert one's intuition about future volatility into current option prices than it is to guess at the appropriate prices themselves.

**Models Are Used to Rank Securities by Value**

Apartments have manifold characteristics and are difficult to compare. Implied price per square foot can be used to rank many similar but not identical apartments. It provides a simple one-dimensional scale on which to begin ranking apartments by value. Implied price per square foot doesn't truly reflect the value of the apartment; it does provide a starting point, after which more qualitative factors must be taken into account.

Given the inevitable unreliability of models and the limited truth or likely falseness of the assumptions they're based on, the best strategy is to use them sparingly and to make as few assumptions as possible when you do. Here are some other rules I've found useful.

## Avoid Axiomatization

Axioms and theorems are suitable for mathematics, but finance is concerned with the world. Every financial axiom is pretty much wrong; the most relevant questions in creating a model are *how wrong* and *in what way*?

## Good Models Are Vulgar in a Sophisticated Way

In physics it pays to drop down deep, several levels below what you can observe, formulate an elegant principle, and then rise back to the surface to work out the observable consequences. Think of Newton, Maxwell, Dirac. Finance lacks deep scientific principles, and so there it's better to stay shallow and use models that have as direct as possible a path between the securities whose prices you know and the security whose value you want to estimate.

Markets are by definition vulgar, and the most useful models are vulgar too, using variables (such as price per square foot) that crowds use to describe the value of the assets they trade. One should build vulgar models in a sophisticated way. Some of the best and most practical models involve interpolation, not in prices but rather in the intuitive variables sophisticated users employ to estimate value, for example, volatility.

Of course over time crowds and markets get smarter, and yesterday's High Dutch becomes tomorrow's patois. The Black-Scholes formula, which translate estimates of volatility into option prices, seemed so arcane when it burst upon the world that Black and Scholes had great difficulty getting their paper accepted for publication. Then, as users of the model grew more experienced, volatility became common currency. Nowadays traders and quants have grown so sophisticated that they talk fluently about models with stochastic volatility, a volatility that is itself volatile.

## Sweep Dirt Under the Rug, but Let Users Know About It

One should be humble in applying mathematics to markets, and be wary of overly ambitious theories. Whenever we make a model of something involving human beings, we are trying to force the ugly stepsister's foot into Cinderella's pretty glass slipper. It doesn't fit without cutting off some essential parts. Financial models, because of their incompleteness, inevitably mask risk. You must start with models but then overlay them with common sense and experience.

The world of markets never matches the ideal circumstances a model assumes. Whenever one uses a model, one should know exactly what has been assumed in its creation and, equally important, exactly what has been swept out of view. A robust model allows a user to qualitatively adjust for those omissions. The Black-Scholes Model *is* robust: its main assumptions are that the risk of an option is related to the risk of the underlying stock, and that the market will be in equilibrium when both option and stock provide the same excess return per unit of risk. That's a sensible idea, no matter how naïvely you define risk. The dangerous part of Black-Scholes is the further assumption that the sole risk of a stock is the risk of diffusion, which isn't true. But the more realistically you can define risk, the better the model will become because the underlying principle is true.

## Use Imagination

The perfect axiom or model doesn't exist, so we have to use imperfect ones intelligently. Smart traders know that you have to combine quantitative models with heuristics. When people build models—when options modelers assume stock prices diffuse like smoke, for example—they make all sorts of imaginative assumptions that they then formulate mathematically.

When someone shows you an economic or financial model that involves mathematics, you should understand that, despite the confident appearance of the equations, what lies beneath is a substrate of great simplification and—only sometimes—great imagination, perhaps even intuition. But you should never forget that even the best financial model can never be truly valid because, despite the fancy mathematics, a model is a toy. No wonder it often breaks down and causes havoc.

## Think of Models as *Gedankenexperiments*

No model is correct, but models can provide immensely helpful ways to estimate value. I like to think of financial models as gedankenexperiments, like those Einstein carried out when he pictured himself surfing a light wave, or Schrödinger when he pictured a macroscopic cat subject to quantum interference. I believe that's the right way to use mathematical models in finance, and the way experienced practitioners do use them. Models are only models, not the thing in itself. Models are better regarded as a collection of parallel thought universes to explore. Each universe should be consistent, but the world of financial concepts and the minds of the humans that interact with them, unlike the world of matter, are going to be infinitely more complex than any model you make of them. When you use a model you are trying to shoehorn the real world into a container too small for it to fit perfectly.

**Beware of Idolatry**

The greatest conceptual danger is idolatry: believing that someone can write down a theory that encapsulates human behavior and thereby free you of the obligation to think for yourself. A model may be entrancing, but no matter how hard you try, you will not be able to breathe life into it. To confuse a model with a theory is to believe that humans obey mathematical rules, and so to invite future disaster.

Financial modelers must therefore compromise. They must decide what small part of the financial world is of greatest current interest to them, describe its key features, and then mock up those features only. A successful financial model must have limited scope and must work with simple analogies. In the end you are trying to rank complex objects by projecting them onto a scale with only a few dimensions.

In physics there may one day be a Theory of Everything; in finance and the social sciences, you have to work hard to have a usable model of anything.

## THE FINANCIAL MODELERS' MANIFESTO

On January 7, 2009, in response to the financial crisis, Paul Wilmott and I published *The Financial Modelers' Manifesto*, an ethical declaration for scientists applying their skills to finance. Here is an excerpt:

### The Modelers' Hippocratic Oath

*I will remember that I didn't make the world, and it doesn't satisfy my equations.*

*Though I will use the models I or others create to boldly estimate value, I will always look over my shoulder and never forget that the model is not the world.*

*I will not be overly impressed by mathematics. I will never sacrifice reality for elegance without explaining to its end users why I have done so.*

*I will not give the people who use my models false comfort about their accuracy.*

*I will make the assumptions and oversights explicit to all who use them.*

*I understand that my work may have enormous effects on society and the economy, many beyond my apprehension.*

## AN ETHICAL COROLLARY

I wrote that modelers of financial securities should make visible the dirt they sweep under the rug. Similarly, the designers of financial products should create securities whose purpose, exposure, and risks are clear. Unnecessarily bundled complex products whose risks are obscure are often more profitable than simple ones because their value is hard to estimate. If products were transparent, good modeling would be easier.

## MARKETS AND MORALS

I am deeply disillusioned by the West's response to the recent financial crisis. Though chance doesn't treat everyone fairly, what makes the intrinsic brutalities of capitalism tolerable is the principle that links risk and return: if you want to have a shot at the up side, you must be willing to suffer the down. In the past few years that principle has been violated. When Wall Street and the U.S. economy were threatened, the ethical principles of capitalism took a backseat. After the bailout of the financial sector, after the provision of cheap government loans to banks at taxpayers' expense, and after the banks' rapid rebound from taxpayer life support to record profits and bonuses, I am ashamed at the hypocrisies of the system. If you want to be in the business of benefiting from the seven fat years, then you must suffer the seven lean years too, even the catastrophically lean ones. We need free markets, but we need them to be principled. As Edward Lucas recently wrote:

> If you believe that capitalism is a system in which money matters more than freedom, you are doomed when people who don't believe in freedom attack using money. Russia has spotted that the weakest link in the Western approach to life is inattention to the moral and ethical basis of capitalism.[3]

## TAT TVAM ASI

In his novel *The Leopard*, Giuseppe di Lampedusa wrote about the effects of the unification of Italy on life and society in Sicily: "If you want things to stay the same, then things will have to change." Capitalism seems incapable of principled self-regulation. What can divert to good purposes the ingenious raw energy of the Will it employs to maintain itself at any cost?

There is no elegant solution. People continue to act out their destiny mechanically, doing what they unthinkingly believe they have to do. Wall Street overshoots in its greed. The Senate committees grandstand. The president seeks to get reelected. The world and the markets silently beg for a Churchill, and society throws them Chamberlains. That's the way of human affairs.

But occasionally there comes that wonderful moment when people who are in a position to make a difference cease to behave mechanically. I think of Mandela and de Klerk, Václav Havel, and Mikhail Gorbachev, men who, rather than fulfilling their preprogrammed destinies, could imagine others as others experience themselves, men who broke the cycle of karma, and so got one step ahead of fate and altered the status quo.

It doesn't happen often, but it does happen sometimes.

# APPENDIX:

# ESCAPING BONDAGE

The title of part 4 of Spinoza's *Ethics* is "Of Human Bondage, or The Strength of the Emotions." Part 5 is entitled "Of the Power of Understanding, or Of Human Freedom." There Spinoza spelled out the plight of people caught in the grip of their passions and his solution to this problem. Following is a diagram with my affective summary of his theory.

# Of Human Bondage
# Of Human Freedom

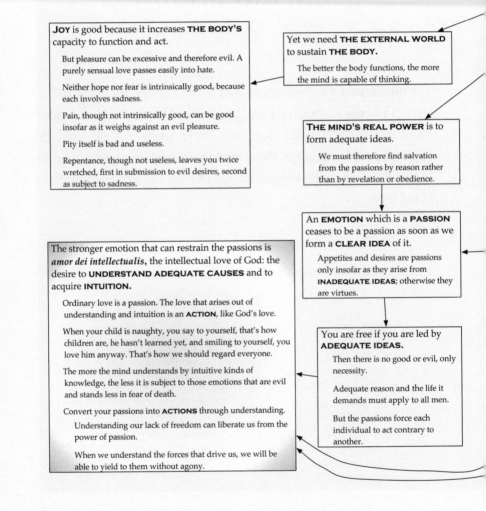

**JOY** is good because it increases **THE BODY'S** capacity to function and act.

But pleasure can be excessive and therefore evil. A purely sensual love passes easily into hate.

Neither hope nor fear is intrinsically good, because each involves sadness.

Pain, though not intrinsically good, can be good insofar as it weighs against an evil pleasure.

Pity itself is bad and useless.

Repentance, though not useless, leaves you twice wretched, first in submission to evil desires, second as subject to sadness.

Yet we need **THE EXTERNAL WORLD** to sustain **THE BODY.**

The better the body functions, the more the mind is capable of thinking.

**THE MIND'S REAL POWER** is to form adequate ideas.

We must therefore find salvation from the passions by reason rather than by revelation or obedience.

An **EMOTION** which is a **PASSION** ceases to be a passion as soon as we form a **CLEAR IDEA** of it.

Appetites and desires are passions only insofar as they arise from **INADEQUATE IDEAS**; otherwise they are virtues.

The stronger emotion that can restrain the passions is *amor dei intellectualis*, the intellectual love of God: the desire to **UNDERSTAND ADEQUATE CAUSES** and to acquire **INTUITION.**

Ordinary love is a passion. The love that arises out of understanding and intuition is an **ACTION**, like God's love.

When your child is naughty, you say to yourself, that's how children are, he hasn't learned yet, and smiling to yourself, you love him anyway. That's how we should regard everyone.

The more the mind understands by intuitive kinds of knowledge, the less it is subject to those emotions that are evil and stands less in fear of death.

Convert your passions into **ACTIONS** through understanding.

Understanding our lack of freedom can liberate us from the power of passion.

When we understand the forces that drive us, we will be able to yield to them without agony.

You are free if you are led by **ADEQUATE IDEAS.**

Then there is no good or evil, only necessity.

Adequate reason and the life it demands must apply to all men.

But the passions force each individual to act contrary to another.

A diagrammatic summary of Spinoza's analysis of bondage and freedom.

Man's lack of **POWER** to moderate and restrain **THE EMOTIONS** leaves him in bondage. How can we find salvation?

**THE EXTERNAL WORLD** produces **EMOTIONS** that can overwhelm and destroy individuals.

An emotion is harmful insofar as it prevents the mind from thinking.

To remain intact, we must understand nature's laws and the constraints that they impose on us.

We feel the greatest **EMOTION** towards someone when we imagine they acted freely rather than of necessity.

If the thing we imagine harmed us freely was in reality not free but acting as a result of laws or causes, we would be less affected.

Nothing is free; we are simply ignorant of most causes.

If we understood all things as necessary, the emotions would have less power. You cannot hate a rock for submitting to gravity.

No **EMOTION** can be restrained except by **A STRONGER EMOTION**.

Men refrain from doing harm out of fear of a greater harm.

The only antidote for emotions is another emotion.

The more **PERFECTION** each thing has, the more it acts and the less it is acted on.

God is entire, complete, perfect.

In a closed system there can be no passive behavior, because there are no outside forces.

There is no separation between the mover and the moved, the creator and the created, the understander and the understood.

If there is no passive behavior, there can be no passion. There are only **ACTIONS** required by **UNDERSTANDING**.

**GOD** is without **PASSIONS**.

He can pass to neither greater nor lesser perfection, and therefore cannot experience pleasure or pain.

Therefore he hates and loves no one.

Therefore it is pointless to try to get him to love you.

Sadness in the world cannot be God's "fault." Once we understand its cause, it ceases to be a passion.

God's knowledge is intuitive, because all he has to know is himself.

The greatest striving of the mind, and its greatest virtue, is understanding by intuition.

Understanding the true nature of things is **THE GREATEST SATISFACTION**. The Dirac equation, for example.

God didn't make the world for humans.

What offends humans is not what offends God.

*To those who ask why God did not so create all men, that they should be governed only by reason, I give no answer but this: because matter was not lacking to him for the creation of every degree of **PERFECTION** from highest to lowest; or, more strictly, because the laws of his nature are so vast, as to suffice for the production of everything conceivable by an infinite intelligence. . . .*

That is how Spinoza puts it. I put it this way: What God did was make everything you can think of. Everything you can think of exists, and everything that exists can be thought of. In Hampshire's phrase, "The possibility cannot be greater than the actual." This is the universe we're in. God made it all, from the apparently imperfect to the **PERFECT**.

# ACKNOWLEDGMENTS

I thank my agents, John and Max Brockman, for their encouragement and support. Without their combination of intellectual interest and practical acumen it would not have been written.

I am very grateful to Hilary Redmon, my editor at Free Press, for her initial faith in the idea that a book about theories and models was worth writing, and for her guidance in shaping it. This book wouldn't have been written without her commitment. I'm also thankful to Sydney Tanigawa of Free Press for her help with the manuscript.

In writing this book I have had the pleasure and benefit of many other people's help. Jeremy Bernstein and Peter Woit contributed useful comments on my account of the development of electromagnetic theory. Dave Edwards did the same, and provided me with additional stimulus in the form of hundreds of articles on topics of diverse interest related to understanding the world around us. Sebastien Bossu read my discussion of the foundations of financial models and gave many suggestions that I incorporated. Richard Cohen, Leo Tilman, and Marc Groz provided useful feedback on parts of the book dealing with finance. Edward Hadas provided insight into the work of Spinoza and related philosophical issues. Nassim Taleb pointed me toward several papers on the philosophy of knowledge and enlightened me about the nature of heuristics. Michael Goodkin lent inspiration and encouragement on several sections of the book. I am also grateful to Wylie O'Sullivan, who helped edit several chapters with great patience and skill. The Santa Fe Institute offered an eclectic, calm, and productive environment for several weeks one August as I was starting to write, and I'm grateful to Doyne Farmer for arranging the Institute's hospitality.

Any defects in accuracy or taste are mine.

# NOTES

## CHAPTER 1: A Foolish Consistency

1. Nonwhites in South Africa at various times were called natives, kaffirs, blacks, Africans, and nonwhites. The term "nonwhites" (*nie blankes* in Afrikaans) contemptuously identifies the salient feature of the black population as their lack of whiteness. I write "contemptuously" because, as I discuss in chapter 3, Spinoza defines contempt as the feeling induced by something whose most salient features are the qualities it lacks. Dirac's theory of the positron, described in chapter 2, also illustrates how absence can become a presence.

2. Normality is what you grow up with. Once, when she was about five years old, we took my daughter on a Caribbean vacation to Anguilla. We stayed in a hotel run by a Mr. Jeremiah Gumbs, a patriarchal-looking black Anguillan who had served as the roving ambassador for his country during its brief 1967 revolution. Mr. Gumbs spoke a lilting biblical-sounding Caribbean English, very different from that of his equally black American-born wife. One day my daughter, who had lived in polyglot New York from birth and was totally comfortable with people of any color, turned to me and asked, "How come Mr. Gumbs is white and his wife is black?"

3. SPF ratings hadn't arrived yet, but had they existed, the sophisticated-smelling pseudo-Swedish iodine-tinted alcohol called Skol that made you look brown as soon as you poured it onto your skin (where it sizzled briefly before it evaporated) would probably have had a rating of around SPF 1/50 rather than 50. According to the definition of SPF, that would have made one minute with Skol on your skin equivalent in terms of facilitating burning to 50 minutes in the sun with no Skol.

4. Hypnosis, as Yuri Manin points out, became possible only after the development of speech.

5. Years later, at a party in New York in the late 1970s, I met a dentist who used hypnosis to anesthetize his patients. I let him take me into a corner of the noisy room and do the arm-raising persuasion on me. Though I "knew" I could resist it, my arm certainly wanted to go up, and so I let it.

6. It is significant that they called themselves Zionist Socialists rather than Socialist Zionists, suggesting that the founders of the party were Socialists first and Zionists second.

7. One of my friends in Habonim, later a judge in South Africa, was euphoniously named Selwyn Selikowitz. My mother was amused at the South African Jewish immigrant parents' propensity for prepending Welsh, Scottish, or Irish Christian names they could barely pronounce to Polish Jewish surnames, resulting in mongrel combinations like Gavin Pasvolsky and Ian Jacobowitz. Years later I discovered how accurately she had perceived the pattern. Whenever in the United States I ran into a Justin or Malcolm Kanopolsky or Rubinowitz, he had invariably been born in South Africa or Rhodesia.

8. Miss Brodie's claim in *The Prime of Miss Jean Brodie,* by Muriel Spark.

### CHAPTER 2: Metaphors, Models, and Theories

1. Schopenhauer combines in one sentence the ideas behind the titles of Henry Luce's three midcentury periodicals, *Time, Life,* and *Fortune.*

2. *Field,* of course, is a metaphor too, the conflation of the qualities of a flattish expanse of grass with the mysterious extension through space of the force that emanates from a massive object. It's hard to talk or write without metaphors.

3. *Arisen* is another gravitational metaphor. I had intended to use a distinct font to indicate every metaphorical word in this chapter, but there are so many that it would severely disturb the readability of the text.

4. I say "describe" rather than "model" for good reasons that will become clearer as this chapter progresses. Deep description is the purpose of a theory.

5. Another fact of nature, the Pauli exclusion principle, prevents two electrons from simultaneously having the same spin and energy.

6. Yehuda Amichai, "The Precision of Pain and the Blurriness of Joy," translated from the Hebrew by Chana Bloch and Chana Kronfeld, *New York Review of Books*, February 18, 1999.

7. You can see analytic continuation at work in the extension of the definition of the factorial function. The factorial of a positive integer $n$ is the formula $n! = n \times (n-1) \times (n-2) \ldots \times 1$ (pronounced "$n$-factorial"), so called because its factors are all the positive numbers less than $n$ itself. When $n$ is 4, for example, then $n! = 4 \times 3 \times 2 \times 1 = 24$. From the definition, it follows that $4! = 4 \times 3!$ and $3! = 3 \times 2!$, and more generally that $n! = n \times (n-1)!$, so that the factorial of an integer is the product of the integer itself and the factorial of the integer just below it. (Describing this integer as lying just *below* another integer is a convenient gravitational metaphor that is almost unavoidable.)

   Using the exclamation point to indicate the factorial is time honored but clumsy. Since $n!$ is a function of the number $n$, it's clearer to reexpress it in terms of the factorial function $F(n)$, defined by $F(n) = (n-1)!$. You can see that $F(n)$ satisfies the oh-so-elegant recursive property $F(n+1) = n \times F(n)$. You can regard this property as a definition of the factorial function, a so-called recursive definition, because it defines the function on the left in terms of the same function (with a different argument) on the right, thus revisiting the function again, "revisiting" being the etymological origin of the word "recursion." If you define $F(1) = 1$, then $F(2), F(3) \ldots$ for all integers $n$ greater than 1 can be found from the recursive definition. You could regard this particular recursion, $F(n+1) = n \times F(n)$, as the essence of the factorial function, as mathematicians do.

   Recursion occurs not only in mathematics, but also in computer science and linguistics. A computer subroutine can recursively call itself, and a grammatical expression is recursive because it can contain other, equally valid grammatical expressions. Recursive functions are interesting because they involve an inner tension: on the one hand they are narcissistically nearsighted, looking only at themselves, local rather than global in scope; on the other hand, because they look only at themselves, they can be more easily extended to vast regions of the globe of numbers.

   The definition $F(n) = (n-1) \times (n-2) \ldots \times 1$ we began with works only for positive integers $n$, because it involves generating a new number by the repetitive subtraction of 1 from $n$ and then multiplying everything that came before by that number. If $n$ is a positive integer and you keep subtracting 1 from it, there is a natural stop at zero. If $n$ is not a positive integer there is no natural stopping point; you can go on forever until $-\infty$.

The definition $F(n + 1) = n \times F(n)$ is more malleable, less obviously restricted to a positive integer argument $n$. The natural question then presents itself to curious mathematicians who always seek to generalize: Why shouldn't there be some other function $F(x)$ that satisfies the relation $F(x + 1) = x \times F(x)$ where $x$ is not necessarily a positive integer but instead some more general number? Why shouldn't the factorial function exist both for $x = 3$ and, say, $x = 3.2731$?

The Swiss mathematician Leonhard Euler discovered (invented?) the gamma function $\Gamma(x)$ that does indeed satisfy $\Gamma(x + 1) = x \times \Gamma(x)$ for all $x$. For integer values of $x$, it agrees with the traditional factorial function. For noninteger or even complex values of $x$, $\Gamma(x)$ serves as a smooth extrapolation or interpolation of the factorial from integer to noninteger arguments. It's smooth because it coincides with the factorial function for positive integer arguments, but maintains the crucial recursive property for noninteger. Mathematicians call this kind of extension *analytic continuation.*

The gamma function $\Gamma(x)$ is a metaphorical extension of the factorial, in which one property, its recursion, becomes its most important feature and serves as the basis for extending it. The extension is a bit like calling an automobile a horseless carriage, preserving the essence of carrying and removing the unnecessary horsefulness, or like calling a railroad *ferrovia* in Italian or *Eisenbahn* in German, focusing on the fact that it's still a road, but one made of iron. Analytic continuation is approximately a mathematical version of synecdoche, a figure of speech in which an entire object is represented by one of its parts, or metonymy, in which an entire object is referred to by one of its properties. Analytic continuation

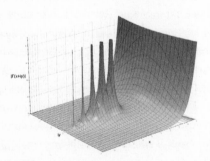

is a method of modeling a function. But whereas most models are restrictive—a model train is less than a real train—in mathematics, a new model can be something greater rather than diminished. That's because mathematics deals entirely with its own world and obeys only its own constraints.

8. How, you should ask, did astronomers determine the true luminosity of Cepheids in order to verify the relationship between absolute luminosity and frequency? There are some nearby Cepheids whose distance can be measured by other, more direct means, and it is from these that one can determine that the true luminosity and the frequency of Cepheids are related. Furthermore, there are physics models that can plausibly explain why frequency and luminosity should be so tightly related, and this sanity check confirms the observations as trustworthy.

9. And therefore more complex.

10. R. P. Crease and C. C. Mann, *The Second Creation*, revised edition (Rutgers University Press, 1996).

11. A model is made. A theory can be discovered. A model always implies a modeler. A theory doesn't necessarily imply a theorizer.

12. By "really" I do mean *really*: every fact is a theory, as are the proton and neutron.

13. Dyson's views on the styles of mathematicians are mentioned in chapter 4.

14. An example of this viewpoint is articulated in *Dictionary of Psychological Medicine: Giving the definition, etymology, and synonyms of the terms used in medical psychology with the symptoms, treatment, and pathology of insanity and the law of lunacy in Great Britain and Ireland*, by Daniel Hack Tuke (P. Blakiston, 1892):

> We are inclined to look upon hypochondriasis from two different points of view; in one we consider that it is rather a form of imperfect evolution and in the other it is a form of nervous dissolution. In the former, the performance of the bodily functions, which should have become so established as to be automatic as far as feeling is concerned, have [*sic*] not developed so far, but remain in the conscious stage, and on the other hand there are conditions of age and disease in which the functions, which have been for years performed unconsciously, again become consciously performed. To take examples, the child learns to use its limbs slowly and with effort, but the time comes when acts of the most complicated kind are performed without any conscious adaptation of means to ends. With age or illness there may be a return to conscious movements.
>
> We believe that many other functions, such as that of digestion for instance, go through a conscious stage to pass to one of healthy

unconsciousness. In one or two cases we have met with patients who have never got over this stage of conscious digestion, and the numbers who in states of ill health or degeneration once more become morbidly conscious of the digestion needs [*sic*] not to be recalled. The constant conscious repetition of sensations, which in the individual or in the race ought to be unconscious, leads to morbid introspection and hypochondriasis; such is our contention.

15. See the discussion in chapter 3 of Spinoza's theory of emotions for an example of a nonmathematical theory.

16. In his Nobel Prize lecture, Feynman wrote,"Many different physical ideas can describe the same physical reality. Thus, classical electrodynamics can be described by a field view, or an action at a distance view, etc. Originally, Maxwell filled space with idler wheels, and Faraday with fields lines, but somehow the Maxwell equations themselves are pristine and independent of the elaboration of words attempting a physical description."

17. Philosophers since antiquity have written extensively about the difference between *episteme* and *techne*. My notion of *theory* is not theory in the sense of "removed from experience" or "the opposite of practice"; my *theory* is analogous to *episteme*, in the sense of knowledge of the real as it is, or *gnosis* in the sense of understanding. *Techne* is closer to craft or heuristic knowledge.

18. "Words are as valuable as money."

19. Italics are mine.

## CHAPTER 3: The Absolute

1. From the Greek *tetra* = four, *gramma* = letters.

2. The ancient Greeks apparently read the Hebrew letters יהוה in the Tetragrammaton as though they were Greek, from left to right, and therefore identified them with the Greek letters they resembled, PIPI, *pi iota*, repeated, and hence pronounced "pipi."

3. "Perfected" meaning not only completed but also made real and perfect, in the Spinozan sense to be discussed later in this chapter: "And God saw every thing that he had made, and, behold, it was very good."

4. Imagine the word *green* always printed in red type.

5. That there is no name for an emotion involving two pleasures could be

taken to validate Schopenhauer's view of the primacy of pain over pleasure. Buddhists, though, do have the term *mudita* to denote a state of joy at the success of others.

6. Spinoza regards clemency as the opposite of cruelty. But though cruelty is a passion we suffer as it overwhelms us from the outside, clemency is not. Clemency, according to Spinoza, is a power whereby man restrains his anger and desire for revenge.

7. In finance an underlier is a simple security, and a derivative is a more complex security whose value depends on the underliers beneath it. A stock is an underlier; an option on a stock is a derivative contract.

8. The contempt for fiat money stems not only from the painless creation of illusory value, but also from its impure nature. It has the soul of an underlier, like the gold that it once laid claim to, but the body of a derivative, a contract printed on paper that, in the end, is anchored to nothing but the government's insistence that it is legal tender for all debts.

9. Somerset Maugham took the title of his novel *Of Human Bondage* from part 4 of Spinoza's *Ethics*, entitled "Of Human Bondage, or The Strength of the Emotions."

10. Gilles Deleuze, *Spinoza: Practical Philosophy* (City Lights Books, 1988).

11. This is perilously close to the modern view that everyone who we would formerly have called bad is now someone's victim rather than a culprit. But as Philip Larkin wrote:

> *They fuck you up, your mum and dad.*
> *They may not mean to, but they do.*
> . . .
> *But they were fucked up in their turn*
> . . .

12. A version of Blake's "To see the world in a grain of sand . . ."

13. The appendix of this book contains my affective map of Spinoza's *Ethics*.

## CHAPTER 4: The Sublime

1. Maxwell's paper "On Governors," published in 1868, created the field of control theory. He used differential equations to explain how to design a governor to stabilize the energy output of a steam engine, so that the engine would automatically slow down when it ran too fast and speed up when it ran too slowly. This kind of cruise control is a desirable trait of economies too, as yet unachievable.
2. A converging field is equivalent to a diverging field with the direction of the arrows reversed.
3. A magnetic pole or charge would be the magnetic analogue of an electric charge. Isolated magnetic poles, a single north or a single south, have been conjectured to exist but have never been discovered. All one ever finds is current loops that behave like north–south pole combinations (Figure 4.2).
4. It puzzles me: What did people think light was before Maxwell identified it with electromagnetic radiation?
5. Quoted in E. T. Bell, *Men of Mathematics* (New York, 1937).
6. Apropos the word *like*, Jeremy Bernstein informs me (private communication) that Marvin Minsky once remarked about something he (Minsky) drew, "This looks *like* a bird, but no bird looks *like* this."
7. Against my best intentions I am writing about the "orbiting" electron. *Orbiting* is a planetary metaphor that fails when one moves from classical to quantum mechanics. There is no "orbiting" electron.
8. The orbit of the electron is also affected by the quanta of its own electromagnetic field, an additional effect I'm ignoring here.

## CHAPTER 5: The Inadequate

1. Hayek also pointed out that the role of psychology is, somewhat paradoxically, to create the path back from laws to the perceptions they explain: "The task of theoretical psychology is the converse one of explaining why these events, which on the basis of their relations to each other can be arranged in a certain (physical) order, manifest a different order in their effect on our senses."
2. To be precise, when you buy a share of stock the first time the company

issues it, in an initial public offering (IPO), the company receives your payment. If you buy a share later, in the secondary market, it's like buying a second hand book; you pay the owner of the book for it, not the publisher.

3. In the long run, of course, paper money is very risky; governments collapse, countries disappear, empires fall, and things hold their value much better than paper does. James Grant, the editor of *Grant's Interest Rate Observer*, has pointed out that an ounce of gold has always more or less been the price of a good men's suit.

4. According to Krugman and Wells, the crisis of 2007–2008 was caused by the global savings glut (www.nybooks.com/articles/archives/2010/sep/30/slump-goes-why/). That was caused by the Asian currency crisis of 1997–1998, which stimulated Asian countries to avoid a repeat by running net surpluses rather than net deficits. But what caused the Asian currency crisis? And what caused whatever caused that? As I pointed out in chapter 3, this is a good example of what Spinoza would call an inadequate explanation.

5. The prevailing price of one incremental share is not necessarily equal to the price per share for acquiring the entire company. With physical goods one usually expects a discount for buying in bulk. With securities one often pays a premium for buying an entire company, probably because the desire to buy an entire company indicates strongly that the buyer thinks it is worth more than it currently trades for.

6. See F. H. Knight, *Risk, Uncertainty, and Profit* (Houghton Mifflin, 1921).

7. See D. A. Freedman and P. B. Stark, "What Is the Chance of an Earthquake?" University of California, Berkeley, Department of Statistics, Technical Report 611, January 2003.

8. In mathematics the symbol $\Delta$ (Delta) before any other symbol indicates an infinitesimally small change in the quantity represented by the second symbol. Thus $\Delta t$ is an arbitrarily small increment in time.

9. See Emanuel Derman, "The Perception of Time, Risk and Return During Periods of Speculation," *Quantitative Finance* 2, no. 4 (2002): 282–96.

10. "Exclusive" in describing an apartment means desirable and expensive because it excludes many people. In a financial crisis, though, exclusivity means illiquidity. What you want to own in a widespread financial crunch are *inclusive* securities.

11. I use the Greek letter omega, pronounced "oh-mega," to quantify deliciousness.

12. F. Black, E. Derman, and W. Toy, "A One-Factor Model of Interest Rates

and Its Application to Treasury Bond Options," *Financial Analysts Journal*, January–February 1990, 33–39.

13. In principle the beta of a stock should be the same forever, its permanent characteristic. In practice, like volatilities, betas change over time. This is one of the problems of CAPM, though by no means its major one.

14. Or, rephrasing the CAPM result in algebra:

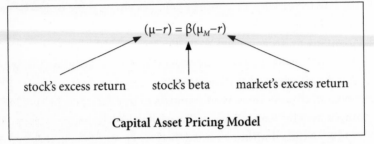

$$(\mu - r) = \beta(\mu_M - r)$$

stock's excess return    stock's beta    market's excess return

**Capital Asset Pricing Model**

The left-hand side of the equation, $\mu - r$, is the excess return you can *expect* to earn when you invest in a stock. The right-hand side, $\mu_M - r$, is the excess return you can *expect* from investing in the entire market $M$.

15. I stress again that CAPM deals with *expected* returns; it is a theorem about what people *should* expect *if* stocks and people behave the way the model assumes they will. Contrast it with Maxwell's equations, which tell you, absolutely, how light *will* behave.

16. When savvy investors find a new way of grouping the market, and that information gets around, other investors will probably begin to anticipate the group behavior, and in so doing accentuate it even more strongly.

## CHAPTER 6: Breaking the Cycle

1. From the Greek word *pragma*, meaning "a material object," pragmamorphism is the inverse of anthropomorphism, which is the attribution of the mental characteristics of human beings to animals or objects. See http://www.edge.org/q2011/q11_3.html#derman.

2. Is a stimulus stimulating when they keep telling you they are going to stimulate you? Do Keynesians laugh when they tickle themselves?

3. Edward Lucas, *The New Cold War: Putin's Russia and the Threat to the West*, 2nd edition (Palgrave Macmillan, 2009).

# INDEX

INDEX

222

# ABOUT THE AUTHOR

Emanuel Derman is head of risk at Prisma Capital Partners and a professor at Columbia University, where he directs the program in financial engineering. He is the author of *My Life as a Quant*, one of *Businessweek*'s top ten books of the year, in which he introduced the quant world to a wide audience.

He was born in South Africa but has spent most of his professional life in New York City, where he has made contributions to several fields. He started out as a theoretical physicist, doing research on unified theories of elementary particle interactions. At AT&T Bell Laboratories in the 1980s he developed programming languages for business modeling. From 1985 to 2002 he worked on Wall Street, running quantitative strategies research groups in fixed income, equities, and risk management, and he was appointed a managing director at Goldman Sachs & Co. in 1997. The financial models he developed there, the Black-Derman-Toy interest rate model and the local volatility model, have become widely used industry standards.

In his 1996 article "Model Risk" Derman pointed out the dangers that inevitably accompany the use of models, a theme he developed in *My Life as a Quant*. Among his many awards and honors, he was named the SunGard/IAFE Financial Engineer of the Year in 2000. He has a PhD in theoretical physics from Columbia University and is the author of numerous articles in elementary particle physics, computer science, and finance.